WHAT ABOUT ME ISM

Vincent F. Hendricks

WHAT ABOUT ME ISM

Insights from Game Theory, Behavioral Economics and Moral Philosophy

Springer

Vincent F. Hendricks
Department of Communication
University of Copenhagen
København S, Denmark

ISBN 978-3-031-84639-7   ISBN 978-3-031-84640-3 (eBook)
https://doi.org/10.1007/978-3-031-84640-3

© The Editor(s) (if applicable) and The Author(s), under exclusive license to Springer Nature Switzerland AG 2025

This work is subject to copyright. All rights are solely and exclusively licensed by the Publisher, whether the whole or part of the material is concerned, specifically the rights of reprinting, reuse of illustrations, recitation, broadcasting, reproduction on microfilms or in any other physical way, and transmission or information storage and retrieval, electronic adaptation, computer software, or by similar or dissimilar methodology now known or hereafter developed.
The use of general descriptive names, registered names, trademarks, service marks, etc. in this publication does not imply, even in the absence of a specific statement, that such names are exempt from the relevant protective laws and regulations and therefore free for general use.
The publisher, the authors and the editors are safe to assume that the advice and information in this book are believed to be true and accurate at the date of publication. Neither the publisher nor the authors or the editors give a warranty, expressed or implied, with respect to the material contained herein or for any errors or omissions that may have been made. The publisher remains neutral with regard to jurisdictional claims in published maps and institutional affiliations.

This Springer imprint is published by the registered company Springer Nature Switzerland AG
The registered company address is: Gewerbestrasse 11, 6330 Cham, Switzerland

If disposing of this product, please recycle the paper.

Photo | Ulrik Jantzen | © 2025

About the Author

Vincent F. Hendricks is Professor of Formal Philosophy at The University of Copenhagen. He was the founder and director of the Center for Information and Bubble Studies (CIBS) funded by the Carlsberg Foundation. A prolific writer he has been awarded several prizes for his research among them *The Elite Research Prize by the Danish Ministry of Science, Technology and Innovation*, *The Roskilde Festival Elite Research Prize*, *Choice Magazine Outstanding Academic Title Award and The Rosenkjær Prize*. He was Editor-in-Chief of *Synthese: An International Journal for Epistemology, Methodology and Philosophy of Science* between 2005-2015.

Hendricks has over the years appeared in a wide variety of media outlets including The Washington Post, The Huffington Post, The Guardian, BBC, Le Monde, DIE ZEIT, Welt, TEDx, The Open Mind:Channel Thirteen, Wired, The Conversation, VICE, Quartz, New Statesman, Frankfurter Allgemeine, Danish Broadcasting Company, News POLSAT Poland, Tom Hartman Show, Pete Dominick Show, The David Packman Show, The Forbidden Speech podcast, SiriusXM, Ideasphere – National Public Radio, Sputnik-News Radio UK, SRF Tagesschau CH, Neue Zuricher Zeitung CH, CNBC AFRICA, New Business Ethipoia, CPH:DOX, TRT World Turkey, ARD Radio Germany , TV2 News Denmark.

Table of Contents

Prologue

1. Life and zero-sum scenarios | 1

2. The whataboutme game | 15

3. Enough about me, what about you, what do you think of me? | 37

4. Moral depravity | 53

5. Seeing it coming a mile away | 71

6. Separate spreadsheets | 87

7. Coward or crook? | 101

8. Work the problem, work together | 117

9. What about you? It's about us! | 133

Acknowledgements

Notes

This one is for you Daddy,
you taught me *some* lesson here ...

Prologue

What-about-me-ism. Another word, yet a new *-ism*, perhaps a neoteric trend – or better still, one more burstable bubble for the academy, online-coaching or self-help literature to take on.

Reasonable sentiment if that's your initial reaction. Whataboutism – or whataboutery – are already established idioms so is this just another spin-off in the same general direction? Whereas whataboutism or whataboutery in the pejorative sense denotes a rhetorical ploy designed to distract attention from a well-placed critical question by retorting to another critical counter-question entailing some counter-accusation against your opponent, whataboutmeism is veritably a very different item. The word has a *"me"* in it. That indeed makes for a different beast altogether.

Whatabout-*me*-ism announces a morbid self-absorption. A view of the world as one mammoth mirror in which one's reflection is not only the object of one's own constant attention and admiration. The whataboutmeist likewise expects everyone else to take pause and admire the mirror image of the protagonist. While at it, with such a main character syndrome, things, events and other people are but props and actors in supporting roles in a movie for which the whataboutmeist is both script writer, director and main character. As a whataboutmeist you are all the above in a production of your own making. Potential failure to realize expectations is not on you, but on the surroundings and the people to whom you have assigned supporting roles whether friends, foes, colleagues, lovers or spouses.

With whataboutmeism as the overarching premise, no matter what goods, advantages or opportunities you have, I am also entitled to. Whatever you gain is indeed something that has passed me by, because why should I not have the same as you? Life is viewed as a zero-sum game. Mere existence accordingly becomes a competition to get what you want. If you don't get it, your loss, either in whole or in part, while the other person's gain: If you get something, it's only because I've been similarly deprived – and net goes to zero.

In the eyes of the whataboutmeist, it is morally unjust: I should be able to achieve the same as you if I have done the same as you – by extension, according to me of course. Therefore, it is now up to you, whether you are the lover, the friend, the surroundings or the world at large, to make sure that I get the same conditions, opportunities, degrees of freedom or rewards as you. And I am morally exempt from responsibility, because on the authority of me, I have done what I could while author of the script, thus factored in is also what is best for us. Now it is up to you, or the world, to fulfill my expectations of, what I think I am entitled to.

In sum, as opposed to whataboutism as a logical fallacy, whataboutmeism captures an ubiquitous attitude towards life and a moral outlook in which:

- Human **deliberation**, **decision** and **action** is
- designed as a **what-about-me game** based on largely your **expectations** towards other people;
- with **no moral responsibility** on your part right or wrong, that's on everybody else, and;
- entails a sense of **entitlement**; if others can have it whether a gift, benefit or reward materially, socially or otherwise, I shall too if for no other reason because … **what about me?**

Some 40 years ago I was given a remote-controlled car by my father and grandmother for my 12th birthday. I adored that Lamborghini-Countach-Cannonball-Run-looking-car. When my friends came over for the birthday party a few days later, some of them wanted to try it out. Time was limited and the cake was about to be cut, so I was not to try it in the same run as some of my buddies. When this became clear to me, I was as puzzled as eventually disappointed and so I asked my daddy: "But what about me?" My father courtly yet crisply explained:

> What about you?! This is not about you! It's about them! You get to play with it all the time. Even if you are not to race it whenever you fancy, cherish the fun your friends are having trying out the RC-car. Oh, and do yourself a favor: Cherish their fun without thinking that you somehow missed out on something along the way just because your friends tried *your* car!

As a kid it took me some time to digest this broadside. Eventually it settled with me, and I have been trying to live my life perpendicular to whataboutmeism ever since – admittedly with varying degrees of success over the ensuing years. Nevertheless, my father's fusillade back then, seems as pertinent as ever today. Designing life as a zero-sum game, whataboutmeism is everywhere to be found offline as well as online where life is now lived as vividly as ever. It's permeating through relationships and love affairs, among colleagues, business partners, family members, relatives, strangers, users on social platforms, friends, foes and freaks.

If practicing whataboutmeism made us more empowered, independent and mature, perhaps such an approach to life could be defended if not condoned. However, it does not, on the contrary. Whataboutmeism must be dismantled, avoided and replaced with something better, which indeed makes us more autonomous and mature. To determine what that is, the epistemological structure and behavioral dynamics of whataboutmeism must be unveiled first. There is a toolbox for doing so including game theory, behavioral economics and moral philosophy.

I recall to this day the sting I felt after my father's barrage relating to the RC car, buddies and the birthday party. There was a certain measure of personal insult, my ego injured, and my expectations were not cashed in neither pertaining to what I felt entitled to, nor what I expected others should immediately realize was to come my way. And amid all my personal presumptions, grievances and self-righteous reasoning, I couldn't help feeling but ashamed of what appeared to be my own initial reaction, inclination and attitude towards the entire spectacle. I felt unfairly overlooked and simultaneously puny.

As powerful, seductive and satisfying whataboutmeism may seem personally, there is also something pompous, predictable and pathetic about life viewed as a zero-sum game hinged on your expectations towards other people. A main character syndrome and a moral trajectory that continuously exonerates you of any wrongdoing since you know best. That's pompous. Squarely placing the blame of mischief on the world and its inhabitants either unwilling or unable to meet with your projected expectations. That's pathetic. And if what-about-me is the way you look at the world, I can see you coming a mile away. You become blatantly predictable.

Suppose everyone decided to live in accordance with whataboutmeism. A world ruled accordingly would be a dire scenario for personal autonomy and collective democracy alike. Autonomy and integrity are not to be confused with perverted egoism as little as democracy is about what society can do for us as we habitually ask: Enough about me, what about you, what do you think of … me? Autonomy is not to be confused with perverted egoism as little as groupthink is tentamount to democracy.

Although having a much tougher time admitting it, whataboutmeism still rears its ugly head from time to time in certain situations, settings and contexts where I take the lead in asking what about me? At least tacitly I may be mumbling these very words to myself when feeling neglected, unappreciated, supercilious or presumably entitled in some way that doesn't come to pass. Why shouldn't I also have a bonus now that my colleague was just awarded one? Why can't I have the same success as an influencer as my neighbor? Why does my good friend spend time with another acquaintance instead of me? Why does my girlfriend choose to spend her night off at the steak house with her friends... what about me?

We can probably most of us relate to just as much. But I'm dead set to fight whataboutmeism all the way because, by the end of the day, it makes me pompous, predictable and pathetic in the eyes of those around me – and by reflection, in the eyes of the beholder. If untamed, whataboutmeism may create a chain reaction among social animals, making us less trusting, less incentivized to coordinate and cooperate. You simply do not put stock in the judgment of others, show neither trust nor desire to cooperate for the good of everyone – yourself included. Taken to the limit, lack of trust, coordination and cooperation is the day of doom for individual autonomy and collective democracy.

To win the personal battle not losing the collective war, it's indispensable for a victorious outcome to understand what whataboutmeism all is about; definition; epistemology, exemplars, behavioral characteristics, structure and dynamics, philosophical underpinnings and moral fabric.

The final analysis is to include tips and tricks of the trade to avoid walking through life, or good portions of it, pompous, predictable and pathetic. "Who would ever want to do that?" is rhetorical enough and not about self-help, coaching, pseudo-psychology or the invention of yet another -ism to the amusement of philosophy and the academy. It's about being good people on a daily basis.

Vincent F. Hendricks
Copenhagen 10 | 01 | 2025

Chapter 1
Life and zero-sum scenarios

Whataboutmeism doesn't only appear in the head of a 12 year old boy being "deprived" of trying out his own RC car at his own birthday party because his own friends are given the chance to race the toy. Whataboutmeism appears micro- and macro-scopically in a bouquet of disparate situations, say, among friends and family members, among married couples, at work, or while trading stock leisurely. Below is a set of real-life samples to the set the scene and prime the pump.

1.1 Friends

Whataboutmeism may be triggered in you largely involuntary and to be sure without any such intent. After a nice dinner out with friends, the waiter comes around with the check. Present is a friend of hours, not known for ever paying a penny too many. This person is exceedingly nitty-gritty about who had what at how much. It is apparently to ensure fair splits of the bill relative to the respective consumptions from food to booze of the parties present. The level of fussy bookkeeping is annoying in and by itself. It doesn't help that after announcing who owns what down to the last nickel and cent, our friend half-jokingly yet seriously proclaims: "Now, I did the arithmetic, I'll leave the tipping to you."

I have no problem with friends not having the necessary financial means for a dinner out. The rest of us would happily have covered that – and then some. But I may have a problem with being cheap while our friend is simultaneously indicating that the self-initialized fair-hair-splitting-arithmetic is such a favor to the rest of us that the party in question is exempted from tipping the waiter. I immediately start thinking, well, I could have decided to be over-particular and done the math too. Enter a nagging feeling of what about me? I'm sure the rest of the circle of friends may entertain similar thoughts. Our mutual friend triggers whataboutmeism in me. I hate myself for taking the bait and possibly becoming something, I truly don't want to be – a fussy friend.

A similar pattern is also sometimes recognizable with high maintenance (and possibly jealous) friends. We are close friends based on reciprocity and respect. However, if you don't call me regularly (although I rarely ever call you), I take it as a sign of you having less of an interest in – and consequently less respect for – me! Given the principles of caring, respect and reciprocity we presumably both have subscribed to, I'm now well within my right to ask you: "What about me"? Now, your high maintenance friend and you may agree on the fundamental theoretical principles of friendship yet disagreeing on the standards of implementation. How often do you call each other, attend each other's parties, how often should you meet for coffee, and so on. But as the high maintenance chums take their standard of implementation to be the yardstick of your sincerity and commitment, you imperceptibly may get caught in a web of conventions and blame-games-of-your-fault which you originally had no intention of signing on to neither in theory nor in practice.

1.2 Family

Families are often thought of as constellations of people caring for each other one way or the other in a household. They come in all sizes and shapes these days from traditional two-by-twos or nuclear families – two parents of two different genders, male and female, with one, or two+ children. Then there are same sex families and rainbow constructions of two parents of the same or multiple genders with twelve kids, two dogs and a turtle named Napoleon. Single-parent families with x kids, four people families with no kids, polyamorous families with some kids, stepfamily, extended family, grandparent family and all sorts of combinations of the above.

No matter the family structure, the family dynamics are given by the duties, obligations, reservations, rights, … all the way down to *expectations* family members have pertaining to themselves and other members of the household. To narrow down a prototypical example of whataboutmeism in a family structure restrict attention now to a-two-parents-opposite gender-stepfamily constellation with kids from previous encounters possibly augmented with a common new offspring.

Partners in second-round stepfamilies come with loads of baggage ranging from previous nuclear experiences, former best practice principles, good and bad memories, abuse or curling care, affections for former extended family members, relations to the ex(s), and again off-springs of past affair(s). A complex with many moving parts times two, one for each of the news partners having decided to try it over again in new surroundings: New villa, old house, rural area, cosmopolitan neighborhood and this latter list is likewise countably infinite.

Now one partner comes with nuclear experiences and unbending principles of family dynamics dating all the way back to own childhood. The other partner comes from, say, a patchwork background with parents and previous partners running in and out of an open and sometimes confusing household governed by mostly workable pragmatics rather than immutable ideology. A question may soon arise: How to bring up the pool of kids brought together intersecting taut principles and prudent tactics while factoring in the wishes, desires and

1.2 Family

demands of the original other parent(s) and extended family entourage?

If one of the partners in the new stepfamily strongly believes that the nuclear model should reign again being a one-model-fits-all, then every new family member should succumb accordingly. No special treatment for anybody. All members of the step- and extended family are equal entries in the same spreadsheet, children ideally count the same independently of where they came from like real siblings, new offspring etc.

Suppose one partner then decides to take the child of a previous relation to the movies, just to have some time alone with the child living mostly with the other biological mother or father. Evidently, that is time spent which on a later date must be made up for either with the current partner and/or other kids to balance out the universal spreadsheet ruling them all evenly intended to make things equal, fair and transparent: Time spent alone with your kid creates a deficit to be covered later. It is a balance sheet of a zero-sum game. The time spent now on one thing must be made up for later because that's the only fair thing to do, the way the weights of the relationship are calibrated.

Of course, the parent deciding to allocate time with the child of past encounters may soon enough feel between a rock and hard place. Damned if you don't spend time with your kid, damned if you do too apparently. Both partners may agree that the relationship should be based on principles of fairness and equality among (new) family members. Yet disagree as to whether the best way to install such principles is by way of designing a zero-game and only ONE universal balance sheet to keep track of all scores no matter the nature, context and circumstances. Be that as it may, any infraction may license either one of the parties to ask: "What about me?" Given the family spreadsheet balance I'm entitled to something too – so I repeat: "What about me?" The same question may be asked by a grandparent annoyed that either grandma or grandpa spend too much time with the grandchildren; competitive relationships between siblings, cousins, uncles, aunts and on it goes for all kinds of family structures ranging from nuclear and rainbow over polyamorous to same sex relations.

1.3 Promotion

Promotions are tricky things to handle. In principle you would like to hand out such to whomever and whenever they are due. And if two are in the running for a promotion you'd often like to award them both – although it is rare one can pull that off. Often you are forced to choose between one and the other. Then it is a zero-sum game for the two parties chosen between. What one wins, the other loses out on – namely the promotion.

1.3 Promotion

I had a situation where two brilliant candidates were in the running for promotion. They thought that there was just one promotion to fight for. So did I at first. Thus, they started competing between them catering to my favor and as such things tend to go, made for an increasingly toxic environment of competition. A zero-sum game environment. Soon enough I realized that if I moved some funds around, balanced out another account I might find the means to reward them both. But before telling them, I called them to my office to inform them, that we compete externally and cooperate internally, and we take pride in each other's victories without envy or what-about-me. I also informed them that either I would find the funds to promote them both or I would have to let them both go. From that moment on, they were thick as thieves and eventually I promoted them both. It boils down to internal competition versus collaboration in the workplace. If two employees compete internally, it may indeed harm the workplace. If they cooperate, it may be to the benefit of both employees and workplace.

But I've also been asked to promote somebody under the heading of apparent entitlement according to the party in question. That didn't go down so well. A position was up for grabs with multiple applicants. It came down to two competitors each with their unique, yet quite different, skillsets both of which would come in handy for an important project running. A project soon to be to be signed, sealed and delivered. Again, I managed to find additional funds in a shoebox in the budget somewhere, hiring them both although with very different job descriptions within the same project.

Soon thereafter, a third party, who had not, but very well could have, applied for the initial position knocked on my door. The employee was frustrated that I had not hired the person in question as *two* positions, not just one, had been filled. This person was very gifted and hardworking as well, but narrowly focusing on an entirely differently field having expressed little to no interest in the project in the pipe to be soon delivered for which the other two parties were hired. But now that they got hired, this third person thought a promotion should be forthcoming too – because, what about me? If they got a break, I should be entitled too, although comfortably subduing, that I never applied for the original job due to lack of interest in the first place. As employer I had to explain the person not to confuse apples with oranges while doing sums in just the ONE spreadsheet. Asking what about me during this calculation, there is only one answer to be had. Chances are you are not going to like it: "What about you? This is not about you!"

1.4 Stock trading

Recreational stock trading is now very easy and quite popular. Stock trading on a leisurely basis has become

1.4 Stock trading

what buying sweepstake tickets and lottery coupons used to be. Trading in stocks for R&R has a rapturous ring to it with an air of prestige surrounding it making for excellent conversation topics at dinner parties among people of all ages with money to spend. While both trading stock and purchasing lottery tickets are about gambling, stock trading is taken to entail some measure of skill rather than just blind luck apparently adding to the social status of stock trading over sweepstakes.

At some dinner function I overheard a discussion between two guests ruminating over their spare time stock portfolios. One had just succeeded in short-selling and was bragging about pocketing the difference after borrowing shares from a broker and then immediately selling them expecting the stock price to fall soon thereafter. And indeed, the stock price had dropped. So now the person in question could buy back the shares at a reduced price, then hand them back to brokage firm and retain the difference as profit. Neat trick befitting an apparently slick mind.

The other party to the conversation was not amused having recently tried the exact same thing though without the success. The difference being that this other party's expectation of the stock price falling had not been cashed in, so no positive difference to be made in the end, just deficit. Complaining about how the market had screwed this person, presumably as able and avid, as the other party to the conversation, somehow it was now the fault and responsibility of the market that things had gone sour. If others can succeed, and I'm at least as good as them according to my own yardstick, possibly even better, then why am I not rewarded accordingly? Obviously, it couldn't be the fault of the other party to the conversation. That leaves the stock market *per se* as the moral culprit because the person in question had done exactly as prescribed. Again, it is simply now unfair that this person doesn't get a similar reward, so it seems only reasonable to ask: "What about me?"

1.5 Traffic

Whataboutmeism may be ignited in each of us, in the meet with demanding or hypocritical friends, in relating to family members and in the workplace. It may also occur when you spend your leisure time trading stock or crypto, and eventually the stock market or crypto exchange is to blame if things go south for you. Now imagine if something as simple as traffic was what about me. So how would we fare?

Road users, whether driving cars, trucks, bikes or bicycles exhibit a self-serving bias when making attributions on the road. When another driver cut us off while switching lanes, we tend to get outraged when making a *dispositional attribution* to the other driver: The one cutting us off is inconsiderate, aggressive,

1.5 Traffic

weaseling in or just plain rude. Now, when it is our turn to cut somebody off, things change to a *situational attribution* favorable to ourselves: "I had no choice, the off-ramp is right there and I have to get on to it" or "I'm really late for the meeting, so I had to jam in".

Now it may be a common fear of road users that they will violate the tacit rules of the road and other drivers will hate them for it. If I'm causing inconvenience to another road user, the rule of thumb is to expect the person to react negatively. The default rule in traffic is that we cut each other no slack and routinely attribute negative reactions to all other road users. Although much of what we imagine others are thinking of us as drivers may indeed be pure fictions, these fantasies are nevertheless standard attributions made in traffic. These figments of our imagination are weaved into the fabric of thoughts and feelings while piloting along.

Causal attributions in traffic are frequently arbitrary making our reasoning about ourselves and other road users biased favoring of our own actions even when we are wrong. We take credit for skillful maneuvering while squeezing into a densely packed line of cars or blasting along faster than usual. But when discovering a parking ticket in the windshield, to blame is the over-zealous metro traffic cop. Congratulating ourselves on a traffic job well done yet criticizing the metropolitan police for exactly carrying out their job in an orderly fashion. This kind of reasoning is biased, self-serving, inconsistent and generally as unpleasant as it is troublesome. A self-serving line of reasoning in traffic is maladaptive and potentially self-destructive. The law of the land seems to be that others are treated harshly by considering their behavior as a sign of what they freely choose to do; yet we excuse what we do ourselves by considering it as "forced" upon us as an inevitable result of the situation. Everyone else is a smug self-serving idiot, and if I make a mistake, it's really everyone else's fault. Maladaptive and destructive.

Imagine on this background that traffic was organized and performed in accordance with what about me. There would be accidents left and right, traffic laws would not have any constitutive value only regulatory use at best. Traffic lights, speed bumps, streets signs and crossings would have but symbolic value. Every human being for himself or herself. Everybody would be shouting what about me as matter of routine while tossing out negative dispositional attributions to other road users yet dressing our own actions up with favorable situational attributions about what we, according to ourselves, were forced to do. If I cut you off, you might miss out on something; if you cut me off I definitely do too. So, there is the zero-sum scenario.

There is good reason why traffic is regulated by external parties to ensure the cooperation of road users because driver's psychology apparently is not going to cut it. Well, at least not without huge losses on all sides possibly ending in self-destruction. It is the same with whataboutmeism. *Cooperation* and *trust* are keys to get out of the pompous, predictable and pathetic position. But before getting ahead of us, nuts, bolts and moving parts of whataboutmeism are to be unveiled first.

Chapter 2
The whataboutme game

The game portrayed in many an action film, in which two cars drive towards each other on collision course at high speed and you win by being the last to flinch, is called the chicken game. The loser is left as a "chicken" by being the first to swerve to the side to avoid being hit by the oncoming car. Designing life as a zero-sum game as a function of your projected expectations is like playing a game of chicken where you expect – and demand – the other side to flinch for the apparent common good defined by … you! Strictly speaking, it is no longer a game of chicken but the *whataboutme game*. With a little luck, you get out having it your way. If not, your antagonist is to blame for both the situational misfit and moral mischief. You may keep playing such games to you reach a war of attrition. By that time, it is all but too late – for you too who set up the loser's game in the first place.

2.1 Tales of Addison and Madison
In chess, when I win, you lose—unless it is a draw where neither of us win nor lose and the game ends in a tie. The same goes for other games like poker and bridge. They are all instances of zero-sum games in which one player's win is equivalent to the other player's loss, so

the net improvement in the game adds up to zero. Thus, the name. The taker of the biggest piece of cake at a birthday party leaves so much less for the rest of the ones to be potentially served and that adds up to a zero-sum game. That is of course if everybody wants a piece and value units of cake in the same way.

There is also the possible outcome of no one getting any cake at all, as we are in the process of tearing each other's heads off in disagreement about how it should divide in first place. Then we all lose, which is why it's technically no longer a zero-sum game. But it does say something about how rare true zero-sum games occur. It also explains why the whataboutme approach as something close to a zero-sum game.

In life, it doesn't have to be the case that one loses exactly what the other wins, but all the same still loses out on something. While at it, we can feel quite differently about winning and losing – sometimes it means little to us, other times a lot – depending on what's at stake. Outcomes may thus be weighted differently by different partners, colleagues, lovers, players. It does not strictly amount to a true zero-sum game either. What stands to reason is that one stands to gain something by getting his or her way, while the other loses something by bowing out. Then one may argue about how much it means to the individual. All the same

2.1 Tales of Addison and Madison

it remains a contest as to who wins and who loses in what-about-me scenarios.

If Addison decides to spend time with the child from a previous marriage and, Madison, the current partner, in consequence feels, that time spent by Addison with the child is time lost which Addison could or should have spent with Madison, then Madison is viewing the entire situation as a zero-sum situation.

Furthermore, Addison and Madison are not only partners, but also colleagues in the same company with the same job description. Now Madison just learned this afternoon while Addison was out with the offspring of previous encounters, that Addison just got a promotion, but Madison didn't. According to Madison, they are equally good at their jobs. Addison only got the promotion because Madison didn't, and again, the total gains added up while subtracting the loses sum to zero. This only adds insult to injury on Madison's scoreboard: That'll be two zero-sum games stacked on top of each other and all on the same day. Bad day already.

While Addison is going to the playground after work with a happy daughter, Madison takes the car home but gets busted for speeding by one of those highway patrol camera traps. Other motorists were speeding too as far as Madison could tell just following the flow of traffic, but somehow Madison got singled out and took the fall for everybody. Their gain, Madison's loss again. Third time is not a charm when life is arranged as a zero-sum game just stacking mischievousness upon misfortune upon misdoing and paying the price while others get away with it clean according to Madison's scorekeeping of events.

And to make matters even worse, Addison and Madison share the common interest of recreational stock-trading and their performance records with respect to their separate portfolios have over time proven Addison and Madison equally avid and able. While on the playground with the child from a past relation, Addison gets a notification that some of the stock in the portfolio just paid off handsomely. What a great day this is! A happy daughter on the playground, a promotion at work and now this! An SMS to Madison is in order arranging champagne cocktails for later celebrating how good a life the two of them have together. On the receiving end however, Addison's partner is getting ready to climb the walls. Madison's stock didn't do anything much today either way, and the fact that some of Addison's stocks just paid off lavishly just leaves Madison feeling, that Addison's win somehow is Madison's loss.

Between Addison spending time with a happy daughter at the playground, the promotion, the speeding ticket on the way home only topped up by Addison's stock gains, there will be no champagne cocktails tonight celebrating togetherness, trust, cooperation and the world at large with Addison, exactly when Madison's outlook and design of life is a zero-sum game.

2.2 Game theory

Game theory is a discipline developed in mathematics, economics, philosophy, psychology, political, social and behavioral science.[1] As the science of strategic thinking it comes with a toolbox for analyzing and understanding a variety of situations where one person's best course of action essentially depends on what other persons do or are expected to do. It is a set of tools for studying interaction whether cooperation or conflict in some environment ranging from dinner parties with family and friends over dealings with colleagues at the workplace to trading floors at a stock exchange, traffic encounters, cold war scenarios and the list goes on.

Game theory may quickly turn exceedingly technical and the mathematical models of strategic interaction among rational agents comparatively complicated. One may however make do with much less while still reasoning out the rational leads to avoid the pompousness, predictability and patheticness of whataboutmeism. Interesting sidenote, game theory was first put to work to analyze two-person zero-sum games which is exactly the current point of departure.

A celebrated and much studied zero-sum scenario is the game of *chicken* which is a model of conflict mirroring some of the listed what-about-me-examples presented so far. As already noted, the game of chicken derives its name from a situation in which two opposing drivers race towards each other on a collision course. If both drivers go straight, both may die in the crash. One must flinch, but if one driver flinches yet the other does not, the one who swerves is the chicken exhibiting cowardice behavior and a chickenshit character.

The drivers could be Addison and Madison from before and their game of chicken may be depicted in the payoff matrix immediately below.

		Madison	
		Flinch	**Straight**
Addison	**Flinch**	Tie, Tie	Lose, Win
	Straight	Win, Lose	Crash, Crash

Addison and Madison have two courses of action respectively "Flinch" and "Straight". The payoffs of these actions relative to what the other decides to do are given in the respective cells. The first argument in a cell entry is Addison's payoff, the second is Madison's. Each one of them would prefer winning over tying, tying over losing and finally losing over crashing.

2.2 Game theory

Now, the loss of flinching is insignificant compared to the crash to happen if neither Addison nor Madison flinch, hence the rational strategy seems to be to flinch before the crash is coming your way. "The hard part about playing chicken is know when to flinch" as actor Scott Glenn recounts in *The Hunt for Red October*. Addison knows this, Madison knows this. This could work to either one's advantage. Because believing that the other is rational and will flinch before the crash is to appear, then stay straight, don't flinch at all, and win to call out the other as the chickenshit. As this goes both ways the situation is very unstable which makes Scott Glenn's observation about the game of chicken ever more pertinent.

Another way to recount this unstable situation in game theoretical terms is to say that there is more than one *Nash equilibrium* to the game of chicken. The equilibrium has taken its name after the American game theorist and Nobel Prize laurate in economics, John Nash (1928-2015), portrayed by yet another actor Russell Crowe in *A Beautiful Mind*. Nash observed that in a state of equilibrium each rational player chooses the best response to the choice of the other player. Each player is doing the best he, she or they can. Hence a *Nash equilibrium* is essentially a strategy pair, one for each player, such that neither player stands to gain anything more by altering their own strategy while the strategy of the opponent remains the same. If that came to pass, it would be so-called *pure strategy* equilibria restricting attention to the two situations in which Addison flinches and Madison doesn't and vice versa.

So, a pure strategy Nash equilibrium happens in situations in which players go for some particular choice with certainty. But many a game doesn't follow a predictable recipe. The fun of Rock-Paper-Scissors as a zero-sum game in which "Scissors" cut "Paper", while "Rock" crushes "Scissors" but "Paper" wraps the "Rock" is exactly its unpredictability of the outcome. The simple Rock-Paper-Scissors game is of interest to game theory since it really has no equilibrium in which players behave in a predictable way. If this was the case, the opponent would take advantage of that, win the game while ruining the fun.

As players attempt unpredictability, the game doesn't come with a pure strategy Nash equilibrium. Although Rock-Paper-Scissors doesn't have a pure strategy Nash equilibrium, it does have what's been labelled a *mixed strategy Nash equilibrium*. Players randomize over the possible pure strategies of showing either "Rock", "Paper" or "Scissors" as the best responses to what the other players do, to form a Nash equilibrium in the first place.

Mixed strategy Nash equilibria may not only be tied to the unpredictability of players' choice, but also come

2.2 Game theory

about when there are multiple pure strategy Nash equilibria, but each player prefers a different equilibrium outcome: Enter Addison and Madison again and their relationship at large. They may each have their individual preference for a certain outcome in the love relationship – namely getting their way each, if they can get away with it. Then the other may usefully deflect. Again, it depends on what the other chooses to do.

		Madison	
		Compromise	Get my way
Addison	**Compromise**	0, 0	-1, 1
	Get my way	1, -1	-1000, -1000

Either Addison will get to add over Madison, or Madison will conversely drive Addison mad by the same token of getting their respective ways while the other gives in: I get it my way and you compromise, score 1 for me, -1 for you; you get your way and I compromise, score 1 for you I end up with losing what you gained, so -1 for me – those are the two pure strategy Nash equilibria. Then there is the case in which we both compromise and get zilch; or … collectively crash trying to get our way you and me both ending up in a real bad place of saying -1000 a piece. In a relationship, the latter is really about divorce. That is a significantly greater loss, than what we stood to gain getting our ways respectively or even tying by both compromising.

One could only speculate, but assuming Addison and Madison would like to work out their differences and stay together, -1000 a piece is again about breaking up. At least if repeated times enough, event for event, outcome after same outcome – compared to what Addison and Madison stand to gain respectively by getting their way or compromising. Pay-offs of 1 or -1 must be accumulated miscellaneous and manifold times to make up for just one situation in which the respective parties get their ways both and lose big.

But the latter situation of breaking up is not a possible equilibrium outcome, if attention is restricted to pure strategy Nash equilibria only. In pure-strategy Nash equilibria, both parties losing (or winning, for that matter) is not a possibility. That is, unless either Alpha or Luca wants a divorce (and thus would win the game by breaking up), this is not a pure-strategy Nash equilibrium.

It is however a *mixed*-strategy Nash equilibrium. As Madison is dead determined to "Get my way", indeed the best response is for Addison is to "Compromise" to salvage the relationship. The thrill of their relationship is really captured by considering the mixed strategy equilibrium in which Addison and Madison routinely randomize between compromising and getting their way. In that case, breaking up is for sure one of the possible equilibrium outcomes. It is not much different

2.2 Game theory

from the game of chicken capturing scenarios of mutually assured destruction of nuclear weapons which just as well may have disaster as a possible outcome.

Madison comes across as somewhat of a control freak. Just revisit the situations reproduced here so far; from the fact that Madison would like to exercise power as to when Addison is to see the child of a previous relation, to who should legitimately gain on the stock market, or whether it is fair to be fined for a traffic offence. But randomization and unpredictability as in a *mixed*-strategy Nash equilibrium seem as remote from control as one could possibly get.

Thus, trying to design life as a zero-sum game where you are guaranteed to win is accordingly some ordeal. Now, if you are Madison known for viewing various events as zero-sum deals so far just taking all the hits as Madison believes, how do you go about securing always getting your way and at the same time stay in the relationship with Addison? There is only one recipe for doing so: Set up the ***whataboutme game*** such that:

1. **The mixed strategy equilibrium in which Addison and Madison randomize between compromising and getting their way is eliminated**
2. **Only the pure strategy Nash equilibria remain in which Madison *can count on***

3. **Addison always choosing to compromise, and consequently**
4. **Madison always triumphs with certainty because there is no other strategy available to Addison in which more is gained than by compromising.**

The whataboutme game matrix could look something like this with the payoff in **bold red** as the interesting one:

		Madison	
		Compromise	**Get my way**
Addison	**Compromise**	0, 0	**2, 4**
	Get my way	1, -1	-1000, -1000

Whenever **Madison obtains payoff 4 by getting away with it, Addison is rewarded 2 by compromising** which is strictly better than obtaining a measly 1 for getting away with it. That's yet better than both compromising scoring 0 for each party and definitely better than a head-on collision resulting from both trying to get their way and -1000-units deficit both ways. Realize that Madison gets what Addison is getting by compromising, 2, plus an additional 2 for getting away with it – double-up! What's not to like if you are Madison while Addison is simultaneously led to believe: *That's as good as it gets*.

2.2 Game theory

Constructing such an outcome seems easy enough, so what's the problem? Addison could choose to follow the same recipe, and then they are back to square one of getting nothing out of their efforts to control. The probate trick here is to force the other person into continuous submission, so that you can always count on getting your way. That is not part of the immediate gameplay as it stands. Some normative component justifying why you always must have it your way has to be part of calling out the chicken and morally blaming somebody else but you for eventual mishaps. Madison, in some ingenious intangible, imperceptible and impalpable way, must obtain and retain the right to define the game, always serve and accordingly win the game whenever, wherever.

Although breaking up has been eradicated from the set of equilibrium outcomes it still stands to reason, that Addison and Madison would lose equally if they were to break up. Madison somehow must convince Addison that always cashing in Madison's expectations is the right thing to do for the good health of the relationship. Madison simply knows better and is justified in setting up the whataboutme game in accordance with the recipe above. Having it my way is always for the common good according to Madison. There is something quite pompous about being of the mind that you constantly know better come what may.

Knowing better likewise exonerates Madison of any moral responsibility were things to go sideways. Since it takes two to tango, and Madison is in the clear knowing better how to do the right thing for the sake of both, the moral responsibility for any wrongdoing is on Addison. Should Addison ever decide to go rogue and not compromise, it would be the dearth for both, but that's all on Addison. So, no blame game to be played either, it is never on Madison, always on Addison. On the other hand, the whataboutmeism from setting up the expectations, apparently knowing what is best for the relationship, to moral exoneration in case of projected misconduct or malpractice is all about Madison.

Setting things up this way, Addison and Madison are doing the best they can, given what the other decides to do. Although there is only one thing to do, only one pure strategy Nash equilibrium of Madison "Get my way" and Addison "Compromise", there is apparently *no regret* either as none of them would realize additional benefits if they were to deviate from their equilibrium strategy. Addison gets 2, Madison 3 and that's as good as it gets. Besides no regret, the entertained Nash equilibrium is also trivially a *rational expectations equilibrium* as there is only one thing to do for each party you can count on. No coordination between players in a real game of chicken, but in this rigged game you can *force* coordination by having it your way as the other always gives in. That's coordination of sorts although a far cry from genuine cooperation in the sense of obtaining greater payoffs jointly by working together than what would have been achieved on an individual basis.

2.3 Either one of us is out of here ... probably

It might not be all that easy in practice to rig a relationship according to the devious whataboutmeism recipe forcing the pure strategy Nash equilibrium using expectations, moral blame on Addison and Madison's exoneration based on apparently knowing what is best. It seems quite unrealistic. After all, we don't always get what we want in a relationship. But if you consider the whataboutme game to be unrealistic, there are two things in particular that are worth investigating:

1. What would things look like in their relationship if based on a mixed strategy Nash equilibrium instead where it is more random who compromises and who gets their way?

2. And while at it, what are the possible equilibrium probabilities also assuming, as hinted throughout the tales of the two lovers so far, that Addison and Madison don't attach the same payoffs to the different outcomes.

2 The whataboutme game

Consider the following game matrix:

		Madison	
		Compromise	Get my way
Addison	Compromise	0, 0	0, 100
	Get my way	80, 0	-20, -50

If Addison and Madison both "Compromise", it's 0 each in payoffs. Addison is less set on "get my way", so it is 80 versus 0 to Madison's "Compromise". But dually, it means much more to Madison to "Get my way" resulting in a payoff of 100, than it does to Addison's converse "Compromise" with 0 payoff. This tendency is mirrored when both "Get my way", which is at greater expense to Madison of -50, than it is to Addison of merely -20. In general, it just means more to Madison to "Get my way" than it does to Addison.

Be that as it may, both Addison and Madison prefer that the other back down so they themselves can enjoy the monopoly of control. That's the real struggle and the mixed strategy Nash equilibrium mirrors the predicament: Neither one of them are ready to give up the fight ending up as the chickenshit, coward or submissive partner. Each has a chance of the dominant position with some probability but not with any certainty. So, what are the equilibrium probabilities that Addison and Madison respectively get their ways?

2.3 Either one of us is out of here … probably

To compute these probabilities in the mixed-strategy Nash equilibrium it should initially be observed that Addison and Madison only randomize between "Compromise" and "Get my way" if indifferent between the two actions. And the only way they would be indifferent would be if the expected payoff of "Compromise" is the same as the expected payoff for "Get my way". If the payoff from either one of the two actions would be higher than the other, it would be stupid not to prefer the action yielding the more profitable outcome and consequently go for that with certainty. Randomization and uncertainty pertaining to which action to perform only enters the equilibrium state when the partners are indifferent as the expected payoff being the same for either action.

Returning to the matrix, if Madison gives in, the expected payoff is 0 no matter what Addison decides to do. But if Madison is to "Get my way", the expected payoff now depends on the probability of Addison getting ditto. Thus, let p denote the probability that Addison is the domineer. If $p = 1/2$, then there is a 50% chance of Addison bossing, if $p = 1$, it is a 100% overshadowing in Addison's favor, while 1-p is probability that Addison gives in.

If Madison subjugates if Addison also tyrannizes, it is a -50 payoff for Madison which happens with probability p. But then again Madison is looking at a 100-unit payoff if

Addison backs down which will happen with probability 1-p. All in all, the expected payoff for

Madison if "get my way" = $-50(p) + 100(1-p)$.

Now, Madison will be indifferent to "Compromise" and "Get my way" assuming that the expected payoffs between the two actions are the same. But, it's already been established that by Madison compromising, the expected payoff is 0, no matter what Addison is fixed on doing so,

$$0 = -50(p) + 100(1-p)$$

Solving for p leaving out the details, Addison's equilibrium chance of "Get my way" is 2/3, while Madison's equilibrium chance of "Get my way" is computed to 4/5 similarly by isolating the probabilities making Addison indifferent to either "Get my way" or "Compromise".

Consulting the probabilities and the matrix jointly, Addison stands to a loss of -20 while tyrannizing if Madison also bosses around which will happen with probability 4/5. On the other hand, Addison is looking at a handsome payoff of 80, if Madison folds and there is a 1/5 probability of that happening. So, for Addison in equilibrium, the payoff expected from folding, that is 0,

2.3 Either one of us is out of here ... probably

is tantamount to the expected payoff being the domineer:

$$0 = -20(4/5) + 80(1/5).$$

With the probabilities of the equilibrium in hand for Addison and Madison, respectively 2/3 and 4/5, it is straightforward to compute the probabilities of every possible outcome of their relationship pending getting my way and compromising for the respective parties:

1. The chance of both **Addison and Madison folding is 1/15** given the probabilities of both of them folding multiplied by each other: 1/3 * 1/5 = 1/15.
2. The chance of **Addison and Madison tyrannizing each other is 8/15** given by the probability of Addison bossing around multiplied by Madison doing the same: 2/3 * 4/5 = 8/15.
3. The chance of **Addison obtaining the relationship monopoly while Madison surrenders to the will of Addison is 2/15** as 2/3 * 1/5 = 2/15.
4. Conversely, the chance of **Madison lording it over while Addison caves in is 4/15** since 4/5 * 1/3 = 4/15.

Probabilities are probabilities and don't add up certainty. Nevertheless, it may still be of some comfort to Madison so gung ho-"get my way" to know that there is twice the chance of succeeding over Addison having the same comparing items 3 and 4. These probabilities are outright insignificant comparted to the 8/15 chance revealed in item 2, that they'll tyrannize each other every day of the week and twice on Sunday at substantial expense to both of them. This particular outcome is altogether likely compared to any of the three other outcomes of their relationship including the situation of them both folding which there is only 1/15 chance.

There is a version of the game in which Addison and Madison both stay and struggle to "Get my way" retaining an option of pulling the plug at a later stage. Such a struggle in the relationship could drag out quite a while accumulating massive loses on both sides over time in a *war of attrition*.

And who would want to be in a relationship characterized by dragging and damaging fights in which the end prize of getting your way by far is outweighed by the accumulated costs for your partner … and for you? Unless of course perhaps you can sit on the high ground by the end of day morally exonerated blaming your partner and possibly everybody and everything else for not cashing in your expectations of what you take to be best for your partner, your relationship, your family and friends, social surroundings and SoMe. Incidentally, we should have divorced SoMe long time ago.

Chapter 3
Enough about me, what about you, what do you think of me?

Whataboutme questions pop up left and right, offline and online – while trout fishing, while watching reality TV and podcasts and checking out the latest news from your favorite influencer. The possibility of continuous comparisons between you and the apparently successful others online coupled with the sentiment that if they can do it, I should be able to too, fuels whataboutmeism. Social platforms, from Facebook over TikTok to Snapchat have turned these features of main character syndrome into parts and parcel of their potent business models.

3.1 My friend gone fishing

One of my oldest and dearest friends, main reader and consultant on this book, Bjarke, told me a true story the other day in which, what initially seemed an involuntary and accidental instance of whataboutmeism, in the end dragged him down hook, line and sinker. Now, Bjarke is an experienced and enthusiastic angler having enjoyed the pleasure of recreational fishing for the past forty-odd years even tuning his own fishing rods in ingenious ways and tying his own topnotch flies for his beloved fly-fishing expeditions. He owns a summerhouse in rural

Sweden providing plenty an opportunity for this continuous and innocent love affair of his. Well, so far innocent.

Only a sport fisherman for sure, albeit an ambitious one of a kind, Bjarke has all the right gear collected meticulously over the years. As of late he has acquired a new app for his smartphone intended to convey information about good fishing grounds for say trout, additionally providing a conference for sharing experiences among anglers, a forum for establishing common knowledge related to best practices and the general tips and tricks of the trade. Very importantly, the app is also intended to serve a more noble purpose. It's a measure for ensuring the sustainability of the overall enterprise such that no fishing grounds are depleted of their precious content which may have dire environmental consequences in both the short and long run for the population of say trout and its biosphere.

An added feature of the app is a message board and connected website on a social platform informing its users of whom apparently are the most successful anglers over the various domains of fishing grounds and species of fish. It's this latter part of the app that eventually becomes the problem for Bjarke. And it is all due to some user profile named Håkon. Of Håkon, a common name in Sweden, Bjarke knows absolutely nothing except Håkon apparently is an enormously

3.1 My friend gone fishing

successful angler of trout on the same recreational basis as Bjarke. Seemingly, much more successful so far than Bjarke has ever been and probably ever will be. Well, at least up until this point if Bjarke has his ways.

Because on this particular occasion Bjarke has decided to follow Håkon's lead and travel to one of the trout grounds in which Håkon has been quite triumphant recently. After trying his luck for hours in these waters without a positive result using his favorite turbo-charged fishing rod and newly tied superfly, a certain sentiment slowly and imperceptibly starts to take hold of Bjarke's mind: What about me? Knowing he should suppress the feeling and get on with his own show, Bjarke tries for yet another hour to no avail. The longer he strives in vain, the stronger the feeling of what about me?! Another fifteen minutes pass and on the last try, the line breaks. As a result, Bjarke (in)voluntarily but completely surrenders to whataboutmeism.

Here is the train of reasoning for succumbing to this pathetic state of mind. Håkon is apparently and avid and able angler, assuming no special advantages or predispositions over Bjarke himself. Thus, by comparison, if Håkon could do it, by default Bjarke should be able to too. Given the fact that Håkon has had success on multiple occasions on this very trout ground, Bjarke should be able to succeed at least this once, and what better time than now as he taken the trouble to travel to this exact this spot? It's Bjarke's time to catch a big one here and now – he seems almost *entitled* to if there is any justice around at all. It appears to be Bjarke's right now to have a big score. But no such luck and by the end of session right before he succumbs, the fishing line breaks, which is downright *unfair* on top of everything else. So, what about me? What do I get – and who is responsible for my failure?

By method of exclusion Bjarke reasons that he couldn't possibly himself be responsible as he, according to a reliable and trusted source did all the right things and by all accounts followed in detail the recipe for successful trout fishing. Bjarke is not a religious man and although Poseidon was one of the Twelve Olympians in ancient Greek religion and myth, god of the sea, storms, earthquakes plus horses and considered one of the most temperamental, capricious and greedy of the bunch of Olympian gods, it seems a little far-fetched for Bjarke to blame Poseidon for his misfortune. That leaves Håkon, the victorious angler, as the potential culprit, but then again Håkon knows even less about Bjarke than Bjarke knows about him, so it seems highly improbable that Håkon somehow has decided to cast a spell on him. Finally, there is the possibility that Mother Nature herself has lined up against him for whatever the reason.

That leaves no immediate culprit but Mother Nature, so then she is to blame and Bjarke is still entitled to catch a

3.1 My friend gone fishing

big fish just like Håkon and have the same victorious experience as anybody else because everything else being equal, they are on par. People or online profiles we don't know anything special about are treated as our equal if for no other reason, because we share the same interest and belong to the same group. Håkon is but a proxy or eigen-variable for what Bjarke equally could and should have had, and by real-time online comparison and update via the app and website, this line of reasoning seems to carry merit as well as validity. According to Bjarke he has simply been unjustly treated by trout-life itself which rightly both nourishes and deserves his harm, anger and disappointment.

Bjarke discounts the possibility that although they are both avid and able, Håkon is just marginally better at flyfishing for trout than Bjarke and consequently more successful. That would hitherto not entitle him to lay claim and right to deserve the same exquisite experiences as Håkon, which Bjarke eagerly and expectantly follows via the online threads of the website.

It could also be that Håkon is full of s*** – and hence not as successful as it seems on the app and website. Then again, Håkon doesn't have to be booming for real as long as Bjarke believes Håkon to be. That'll be a good enough propellant for whataboutmeism on Bjarke's part. True or false on Håkon's account may either way still license feeling wrongfully righteous on Bjarke's account.

In his utter dismay, Bjarke gives it yet another shot for which he gets nothing. He does realize however that all the way through this excruciating session there were no forerunners attached to his marvelously constructed superfly in the first place. That'll do it! It was after all Mother Nature that had lined up against him; to be even more fair and balanced, it was Bjarke lined up against himself with a self-made defect fly. Really, he is himself to blame for the apparent injustice, but you may still – as demonstrated – ask what about me all the same.

Much ado about nothing and all this internal commotion and commiserating since Bjarke can compare himself with others continuously online where everybody ingroup may seem equal if for no other reason because they enjoy the privilege of having a social media profile for which some have an apparently desirable online identity. Some of the biggest companies in the world today have turned comparison and online identity into one of the most profitable business adventures of all time. Social platforms have gone fishing in our need for equality, social comparison, opportunity, recognition, projection and expectation stimulating whataboutmeism wonderfully and worryingly.

3.2 Selfie Society

"Little mirror on the wall, who's the fairest of them all?" Every day the queen asks the same question of her magical mirror and hopes the mirror will promptly answer back: "Thou, O Queen, art the fairest in the land." In the digital era we all have a magic mirror, and we all compare our beauty and popularity to the rest of the kingdom subjects. Social platforms like Instagram, Facebook, TikTok, Twitter and Snapchat are types of mirrors or exchanges measuring our popularity by the numbers of 'thumbs up', hearts, followers, views, comments and other forms of 'engagement'. The big difference between the queen's magic mirror and the social platform's mirror is that in the fairytale of Snow White, it is solely the queen receiving an answer. On social platforms the answer is often freely available within the network.

Social platforms are built with functionality that make it as smooth as it is simple for users to size up each other's popularity. When we post on our profiles, payoff comes in the form of likes and comments – or angry faces, downvotes and diligent dismissals for that matter. The apparent popularity is immediate for others to see in terms of likes and followers on the *emoji exchange*, as platform algorithms reward content with many likes by featuring it as popular content to other users.

Take a selfie for instance. A new communicative norm that has taken humanity by storm. A decade ago, it seemed way too vain to take pictures of yourself in public and flash it to friends and acquaintances from near and far — "look at me!". Today taking and sharing selfies with the outside world is not only about as normal as it can possibly be to drink tea. It also signals a fundamental change in social interaction over a very short period that notably coincides with the successful merger of smartphones and social platforms to create an optimal framework for disseminating selfies. In hindsight, and to put things into perspective, it was only in 2013, that the word "selfie" entered our colloquial vocabulary and became the *Oxford Dictionary* word of the year.

The selfie is a measurable product in the digital information market, where attention is the desired return. On social platforms, users of all ages, shapes and sizes share selfies to get a response. There is the 'duckface', also known as pouty lips, which accentuates the cheek bones, the 'fish-gabe' with slightly open lips so the teeth barely show and send a seductive signal, or the completely natural smile #withoutfilter. Like fashion phenomena, the standards for selfies develop over time and become markers in different cultures and subcultures.

Regardless of culture, the premise is the same: A selfie is a unit the value of which may be negotiated and exchanged by users— and that has caught the eye of politicians. If politicians share a political message, for example on Instagram, the selfie is an investment object, that brings added attention to their message — a selfie in exchange for the attention of potential voters; a personal, spontaneous, direct, unfiltered and authentic "digital handshake" between politicians and their constituents. Now, a selfie is trivially a witness to your role as the main character in your own shot.

3.3 Main character syndrome

In 1964, Canadian media theorist Marshall McLuhan proclaimed *the medium is the message.* With this famous line he conceived that a given medium influences the surrounding society and literally creates new types of environments. With cell phones in pocket and profiles on social media platforms we have stepped into an age where we might add: *Man is the medium*. YouTube encapsulates this very idea in its slogan: "Broadcast Yourself," or as Instagram notes, "Capture and Share the World's Moments".

There is no longer an editor to prevent you from going live or being broadcast, as long as you remain within the bounds of the official and ever-changing platform's community standards or guidelines. To be one's own medium is thoroughly a form of strategic communication, where users become homespun marketing consultants that speculate in virality — *What will get the most attention? A selfie, a sunset or bare skin?*

Håkon can do it. Reality stars like the Kardashians can do it, scores of influencers can do it, why shouldn't I be able to make it too as a medium – what about me? I'm the main character in my own show, so why not!?

Back in 1998, key elements of what is now known as the *main character syndrome* were portrayed in the popular TV series *Sex and the City*, episode 16, "Ring a Ding Ding". The main character Carrie Bradshaw (played by Sarah Jessica Parker) oscillates between an evenly lovable and approachable person and what appears to be a narcissistic sociopath. In the episode, Carrie Bradshaw breaks up with her boyfriend in a despicable way, tyrannizes a friend in a way similar when Carrie doesn't get what exactly her main character prescribes. Although the character Carrie Bradshaw trivially has the leading role in *Sex and the City*, it is the way in which the character behaves that is striking. *It is as if she well knows that she is starring in her own show*.

The word "syndrome" that appears in "main character syndrome" is not technically a "syndrome" in the medical sense as a psychiatric diagnosis. It is a concept subject of considerable attention online and not least on social platforms in recent years. Generation Z has

3.3 Main character syndrome

popularized the concept on TikTok. Most often as a sardonic and ironic comment on how certain users behave as if their reality is something out of TV series or movie starring themselves. The reported behavior of the TikTok users in question seems to indicate that whatever is happening is only happening to further their own story. All props like physical remedies or other people appearing in the TikTok videos are only there to promote the particular user's story. You yourself is the protagonist. Everyone else is reduced to supporting roles of either helpful hands or disgusting villains. These characters are included at appropriate places as part and parcel of a plot scripted by the performing TikTok'er. As the protagonist, you typically portray yourself as the ultimate hero or the corresponding victim. Not only TikTok is used as a channel; Facebook, Instagram, YouTube and the other social platforms may also serve the purpose at hand.

If you routinely view yourself through the narrator's lens, taking center stage of your own story, the way you present yourself and how you behave in public naturally changes. When a friend confides in me, my immediate reaction may be a romanticization of my own problems, or a downright rejection of other people's problems. If others have problems while I have the main character syndrome, such problems are ultimately about … me. With a main character syndrome, one may be invited to a birthday party to celebrate someone else's birthday.

Still, most of the time is spent taking selfies and posting about one's own experiences at the other's birthday. Or commiserating to your father about your friends wanting to try out your RC car when you turn 12.

In relationships, the main role syndrome may of course also give rise to certain complications. Constantly focusing on your own feelings, needs and achievements, there is little room for the emotional universe of your partner. If a little room is granted at all, it is merely as means in a narrative about one's own feelings, needs and aspirations. If two people with each their main character syndrome get together in a romantic relationship, it may very well end up with two different plots. One is but a prop in the other's story and vice versa. Two completely different productions, each with their respective audiences. Two protagonists not approaching each other as anything but romantic window-dressing. The danger is that you may indeed be the hero in your own story while the villain in your partner's movie. Both may be true at the same time.

The main character syndrome thus refers to a series of thoughts and behavioral patterns that may arise when a person basically imagines himself or herself as *the* central role in a possibly fictional version of his or her own life.

Now, sounds suspiciously like something that could

spark whataboutmeism online given the constant social comparison to friends and foes in a network.

3.4 What about me ... SoMe

The OECD has defined social capital as "networks together with shared norms, values and understandings that facilitate co-operation within or among groups". From here, social capital is divided into various forms including *bonds* between agents with common identity, *bridges* to more peripheral relations immediately transcending common identity like colleagues and associates or acquaintances and *linkages* to people or groups either further up or down the social ladder.[2]

To trade social capital through OECD's bonds, bridges and linkages is like trading in hard currency. We only accept a currency we know others will accept as well. So user profiles look much the same, imitating more or less what seems apparently successful for influencers and other traders and high-rollers of the social casino with scores of followers; everybody use the same accepted templates for photos, videos and memes, users play the same popular games, follow the same distinguished profiles, have the same in-group opinions about such and so, subscribe to the same sites, channels, outlets for beauty products and lifestyle advice and the list goes on.

The social influence on social platforms is so massive as it has become much easier to follow the movement on trading floors for the apparently successful life. We compare ourselves with other in the network and in doing so we are constantly in front of the mirror reflecting the life of the other taking stock of how well we measure up to that very reflection. The mirror image is an integral part of establishing the relational "I" – the place of the subjective being in kinship to other knowns as well as unknowns. The relational "I" is borne by comparisons – who am I in relation to you and others? On social platforms of scale, the social influence may become so potent and pervasive that the singular subject is exiled backstage, while the relational "I" is frontstage constantly assessing and appraising itself with reference to other in-group members, their looks, appearances, opinions, interests and popularity.

There is a distinct danger in this constant process of comparison and juxtaposition as noted way back by the Danish philosopher, Søren Kierkegaard (1813-1855), in his seminal treatise on the human condition, *Either/Or* from 1843. The relational "I" bids us to constantly wear the mask, by which we lose our *self*, becoming a *nonentity* as Kierkegaard calls it:

> Your occupation consists in preserving your hiding place, and you are successful, for your mask is the most puzzling of all; in other words, you are a

3.4 What about me ... SoMe

nonentity and are only something in relation to others, and what you are you are only through this relation.[3]

Being rich on social capital and enjoying social recognition are very human traits. It is in part our need for being seen, reckoned with and recognized which social platforms have turned into an integral part of their profitable business model. Social platforms strive on our expectations for a reward: Enough about me, what about you, what do you think of me? No wonder SoMe fuels the bonfire of the vanities and whataboutmeism and its perverted moral code, rather moral depravity.

Chapter 4
Moral depravity

The moral philosophy underlying whataboutmeism is a perverted moral position or conniving cocktail obtained by turning Immanuel Kant on his head and mixing this composite up with elements from ethical egoism. Then by projecting your expectations for others to realize in the aim of the greater good defined by you, you are systematically exonerated from any moral responsibility and accountability. Moral responsibility and accountability are lopsidedly placed on social surroundings in which what about me is practiced. That's a new position in moral philosophy not encountered nor characterized before. What does it look like exactly?

4.1 It can't be Kant

If you were to ask average humans whether they consider themselves dignified, autonomous and responsible beings, the default answer is probably going to be a reassuring yes! Yes, by and large, I'm an independent, free and responsible agent making sound decisions based on available information about which chest freezer to by next and who should be voted into public office. Now, you may try to insult people in a million ways from Sunday, but if you attempt to insult them on their intelligence they tend to get pissed-off something awful.

We may not be moral beings by nature, but a moral framework surely may be designed as to how we should act and that's what normative theories of morality are all about – Kant's included.

Dignity, autonomy, responsibility and reason are cornerstones in the moral philosophy of German philosopher Immanuel Kant (1724-1804). As one of the most formidable thinkers of the Enlightenment, Kant thought that the morality of man hangs from the garland of reason. He proposed that all humans are equal and possess innate dignity by virtue of being *moral* agents— and humans alone are moral agents. This very dignity carries certain ethical implications. What Kant calls the "transcendent kernel" of man grace all humans with unconditional, intrinsic worth and should treat each other accordingly.

To this day, these philosophical ideas and ideals still carry substantial weight as to our common view of how the world should be. By way of example, The United Nations: Universal Declaration of Human Rights, Article 1 reads:

4.1 It can't be Kant

> All human beings are born free and equal in dignity and rights. They are endowed with reason and conscience and should act towards one another in a spirit of brotherhood.[4]

There is a strong Kantian flavor to this first article of the UN charter – and for good reason too. Dignity, intrinsic worth and moral obligations should lead us to reason that whatever we decide to do individually, which has ethical implications of which we should be conscience, should generalize to hold for all fellow humans as well. That's how the "brotherhood" of Article 1 is instilled and maintained: If I adhere to some code of conduct, I should wish and will that it may hold for everybody else too – that's exactly what makes us brothers and sisters as a species. If not, I should refrain from pursuing the set of actions in question. That's the message of Kant's famous categorical imperative in his moral philosophy:

Kant's categorical imperative
"Act only according to that maxim whereby you can, at the same time, will that it should become a universal law."[5]

Now, it may be tempting to think that all this mention of me, dignity and autonomy should be the ideal points of departure for determining the moral underpinnings of whataboutmeism. But really, whataboutmeism is Kant's categorical imperative turned on its head: I expect you to act in my best interest, but I'm simultaneously excused for whatever moral mischief may come along. That's on you as my expectations are busted, but then again, that's your fault:

> **Whataboutmeism imperative**
> *"Act only according to that maxim whereby you can, at the same time, will that others shall realize your expectations."*

In whataboutmeism, there is no moral responsibility attached to its protagonists as that very feature is projected onto the surrounding agents and circumstances by virtue of the expectations put forth by the individual torchbearer in question. That's pretty much as perpendicular to Kant's original ideas as one may possibly turn. It becomes the responsibility of the surroundings to fulfill the expectations of the whataboutmeist. Beneath the surface there is an implicit assumption that 'I' knows what is best for everyone. If one's expectations are thus not fulfilled, it is ultimately the fault of others and thus their possible moral headache. How it's possible to waive moral responsibility will become clear before long. However, it requires a somewhat long-winded argument that rests on a few assumptions to be addressed first.

4.1 It can't be Kant

In Kant's moral philosophy it follows, that the dignity of human as moral beings coupled with the categorical imperative again should lead us to *reason* that to keep the brotherhood intact, untarnished and equal among free moral agents, humans are always to be treated as the end goal, and may never serve as but means to some persons' end:

Kant's "Formula of Humanity"
"So act that you treat humanity, whether in your own person or in the person of any other, always at the same time as an end, never merely as a means."[6]

Kant's formula of humanity immediately shines through in the Universal Declaration of Human Rights of the UN. Especially in Article 4, which proclaims:

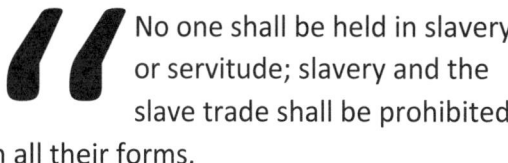

No one shall be held in slavery or servitude; slavery and the slave trade shall be prohibited in all their forms.

But in whataboutmeism, other humans may very well be held in – if not slavery – then at least servitude if it goes to satisfy my expectations and ends, the satisfaction of which is for you to realize, thus your moral responsibility while I'm in the clear either way. That's the formula of whataboutmeism again paraphrased over Kant's original formular of humanity:

> **Formula of Whataboutmeism**
> *"So act that you treat humanity, in the person of any other, always at the same time as a means, to merely your ends."*

Thus, it can't be Kant over which the moral philosophy of whataboutmeism may be adequately molded. It must be much more ego- and self-absorption driven where it may be morally acceptable for humans to be thought of as means ever so often, or always, if need be. And need be in whataboutmeism when personal expectations are to be cashed in, and your ends realized by others in a life designed as a whataboutme game.

4.2 Perverted egoism

Humans seem to act as if they are largely just selfish and thoroughly self-interested either all the time or often enough to try turning it into a moral theory *per se*. Psychological egoism, as a way of describing human conduct was already entertained by the British philosopher, economist and political reformer, Jeremy Bentham (1748-1832):

4.2 Perverted egoism

> On the occasion of every act he exercises, every human being is led to pursue that line of conduct which, according to his view of the case, taken by him at the moment, will be in the highest degree contributory to his own greatest happiness.[7]

For each person there is only one end to pursue – own happiness, welfare, utility or self-interest. Whether happiness, welfare or self-interest may be used interchangeably or come with subtle differences is a discussion reserved for a rainy day. Let Bentham's quote stand, psychological egoism is but a descriptive position, not a prescriptive one yet.

Ethical egoism amounts to the normative theory that you *morally* ought to perform some action or the other only insofar, and because, performing the action in question maximizes your self-interest. Maximizing your self-interest and expected payoffs is your only moral duty.

Ethical egoism seems like a more befitting theoretical point of departure for whataboutmeism than say Kantian ethics, but with an interesting twist to it. While Kantian ethics has it baked into its fabric that a moral agent gives weight to the interest of others, ethical egoism, doesn't immediately come with such a feature but may show up as a derivative. To obtain certain goods, say defense or even friendship or a loving relationship, acting as if no weight is given to the interests of others, chances are that these very others will decide not to cooperate with me, which means I can forget about acquiring these goods of self-interest. Constantly breaking your promises whenever it maximizes self-interest likely implies that no one will accept your promises, will deny cooperating with you maybe even attack in return, the latter of which is not a way of maximizing self-interest. The egoist will do best by acting as if others have weight too – of course provided the others act as if I carry weight as well. In principle, many of the duties that other moral theories come with by default, may show up as derivatives in ethical egoism. Even an egoist may take moral responsibility and exercise duties if none other, for their own benefit. One response while reading through Bentham's characterization of egoism, is really, that's his position? That's tough to swallow in terms of describing human's moral sentiments and behavior. But whataboutmeism is even worse and we ever so often swallow it no problem at least if being on the defining end of things.

Here within lies the convolution with whataboutmeism. The twist is that while egoism realizes it may be in the best interest of the subject to act as if others have weight to obtain certain goods, benefits or payoffs whataboutmeism rejects that premise. The whataboutmeist – who, not to forget, could be any one of us once in a blue moon, ever so often, or all the time – would like to have one's cake and eat it too: To obtain the benefits

4.2 Perverted egoism

of cooperation without granting that others really have weight; and if they do, it is only because they serve as means to realizing your ends as the formula of whataboutmeism dictates. A whataboutmeist will take no moral responsibility, they are beyond such.

To realize this position of perversion recall that the imperative of whataboutmeism bids that one should: "Act only according to that maxim whereby you can, at the same time, will that others shall realize your expectations." Questions:

- Why should others want to realize your expectations?
- Put differently, why should others always accept the pure strategy Nash equilibrium in which they compromise, and you get your way like in Addison and Madison's whataboutme game in their relationship?
- Why would caving in always be in the best self-interest of the other and what about the moral responsibility and accountability if things were to go south?

The simple answer is: As a whataboutmeist, you know better than any of your antagonists how to secure happiness and, just as importantly, *avoid suffering and pain* for yourself and the ones around you independently of whether they believe it to be so. There is even a religious and philosophical argument to rehearse that will work in your favor. The argument has been standing since the days of old.

4.3 The Book of Job is Madison's recipe

The rather pessimistic German philosopher Arthur Schopenhauer (1778–1860) had an unpleasantly sharp eye for how man is a true master of suffering. According to him, our whole life is suffering in different forms. We may suffer in many ways: From basic physical pain and agony, thirst and hunger, sickness, and poverty to lack of recognition, appreciation, fear, alienation, rootlessness, social or political marginalization, and stigmatization, to name a few. Suffering takes many forms, and the question: "Why do I suffer?" is a basic, existential question that has been posed time and again, ever since Job did it in the "Book of Job" in the *Old Testament*.

The story of Job is a story of suffering. The innocent Job is hit by yet one catastrophy after another because God and the Devil have made a bet as to whether Job will stick to his belief in God regardless of massive misfortune. Job's friends are not exactly great friends; they insist that Job must have sinned in order to be hit by such suffering. As they see it, all suffering is God's punishment for sinning, and since God is justice incarnate, the punishment must be just:

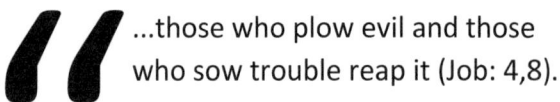
...those who plow evil and those who sow trouble reap it (Job: 4,8).

4.3 The Book of Job is Madison's recipe

They therefore believe that Job must necessarily be guilty but might have forgotten it himself, because you reap what you sow.

Until modern times, this was religion's standard response: Have you been hit by the plague? Surely, you must have sinned. An earthquake? God's punishment! That will teach you. This is the theological *theodicy* tradition, which defends God as good and almighty despite the existence of suffering. The term itself, *theodicy*, is a compound of *teo* = God and *diké* = justice. It was introduced by the German philosopher Gottfried Wilhelm Leibniz (1646–1716), who added the philosophical remark that suffering is a necessary evil in the best of all possible worlds to the traditional religious explanation of suffering as punishment. [8] Supposedly, the reason man cannot see this is due to our limited perspective; but God sees it, and the Lord works in mysterious ways.

There are examples of more modern responses in classic literature. The French philosopher François-Marie Arouet (1694-1778), best known by his *nom de plume*, Voltaire's book *Candide* is a satirical debunking of the perception of suffering as deserved or a necessary evil. In Voltaire's opinion, it adds insult to injury to preach that it is people's own fault, or that suffering is a necessary ingredient in the best of all possible worlds. He rejects the idea that it does not suffice to suffer, but rather posits that one must also endure additional suffering from knowing that one has brought it upon oneself or accept one's own and other people's suffering as building blocks in universal harmony. Russian author Fyodor Dostoyevsky's character Ivan Karamasov in his passionate philosophical novel, *The Brothers Karamazov* from 1880, likewise rejects salvation for that exact reason: If the suffering, particularly that of children, is necessary for the salvation and the harmony of the whole, then he feels the ticket to Paradise comes at too high a price and says no thanks.

Even though the great religious theodicy explanations have gone out of fashion, without however disappearing from fundamentalist circles, the explanatory scaffold has lived on. It has been secularized and has found its way into politics and economics. The narratives that are being served to people stuck at the lower end of the income scale, or who have lost their jobs because of globalization, rationalization, and automation, are akin to the traditional theodicy explanations in their basic structure. Here are a few examples taken to their absurd extreme:

You suffer (say by losing your job) because:

- You fail to be flexible enough for the globalized market,
- You have no useful training or education, or

4.3 The Book of Job is Madison's recipe

- You are part of the "basket of deplorables,"[9] which was how Secretary Hillary Clinton referred to a large part of Trump's electoral base during the 2016 US presidential run.

And this is why you largely deserve the situation that you find yourself in: You reap what you sow. Even if you are one of the people who lost their jobs, globalization is generally a good thing, since it is good for the economy and growth in the long run. Your unemployment is a necessary evil in the best of all possible economic systems, which is why things only look dark from your limited perspective. Seen from above, it is for the common good in the long run and will create growth, jobs, and progress.

Similar theodicy sentiments of explanation may be found in recent economic theory. In the best of all possible economic systems the individual is rewarded according to its contribution and merits. The market is rational because the reward is proportional to effort: the income equals productivity which equals benefit for society. The wealthiest as well as the individuals placed at the bottom of the income scale harvest as they have sown. The growing inequality and the wage stagnation for the middle class is within this reasoning nothing else but a manifestation of a higher justice, of "just deserts".[10]

4.4 Moral position of magic

Whataboutmeism may use the same explanatory blueprint to justify how its proponents are entitled to have their way. If believing that you are constantly being set up for a zero-sum, like Addison does, in the relationship with Madison,

- for which Addison is always to **compromise by meeting Madison's expectation**
- forcing **a pure strategy Nash equilibrium for Madison to "Get my way"**,
- this is only "just" as **Madison simply *knows better*** – not unlike God in the traditional religious versions of the theodicy –
- how **Addison and Madison together gain happiness while avoiding suffering in their relationship**.

Surely, it may indeed feel to Addison as if tyrannized but that's only from the subjected subject's limited perspective. Madison projecting an expectation for Addison to meet in their zero-sum arrangement is a necessary evil in the best of all possible worlds. Could Addison only see the situation from Madison's privileged vantage point, it would be obvious, that Madison's "get my way" is the only way. Madison's tyranny for Addison to take, is but a relationship of a "just desert".

4.4 Moral position of magic

Since you reap what you sow, should things go sideways in their relationship, Addison is to carry the burden and blame. The moral responsibility only has one addressee, and it sure isn't Madison. While Addison, according to Madison's theodicy, carries all the moral responsibility if Madison expectations are not met, for Madison there is no moral accountability whatsoever. Knowing better what the common good is and projecting expectations accordingly absolve Madison from any broadsides of moral blame in case of failure. It is but a consequence of the secular version of theodicy – you wouldn't blame God for any moral mischief as God is good and almighty despite the existence of suffering. Similarly for Madison over Addison – you and me both.

What the common good is taken to be is a function of the situation or context. The common good is what from the vantage point of the one exercising whataboutmeism, is deemed socially most beneficial to the group or setting. In a relationship, a common good agreed to by both parties could be general trust in each other. But the whataboutmeist of the relation, then gets to instrumentalize this trust by defining, for instance, certain "mutually" beneficial rules and regulations as to when, how and where the other party should spend time with the offspring of a previous affair. And it is for the other party to follow these directives, or realize the expectations put forth by the advocate of whataboutmeism, if the common good of trust is to be realized. If

Addison follows Madison's defined template for when Addison should spend time with the child of prior encounters, it is best for Addison, Madison, the continuous trust between them in their relationship and thus the common good they apparently both agreed on in the first place. All theoretically cooked up and practically arranged by Madison but expected executed by Addison. If not accordingly put into effect, or failing along the way, Addison is hoodwinked into taking the moral blame for any detriments, Madison is exonerated by "knowing better" by way of the secular theodicy argument. It is a closed self-reinforcing downward spiraling moral system; the more it is practiced, the stronger it becomes and even more right it may seem to the whataboutmeist constantly tyrannizing the surrounddings.

Among friends, the common good defined or expected by the whataboutmeist of the bunch, could be a particular way of spending time together at a social function, playing a certain card game, enjoying a hide and seek session where everybody is to find … guess who?; singing songs marked as particularly stimulating for in-group cohesion or family consolidation by the one who knows better. Similarly for situations at the workplace, at the playground among parents one of whom is a whataboutmeist, at PT meetings, at a board meeting of the local tennis club and the list goes *ad infinitum*.

4.4 Moral position of magic

Since the common good is a function of the situation or context, there is an *arbitrariness* as to what may be viewed as a common good while being on the practicing end of what about me. I don't have a summer residence, but I'm sure it would take me little effort to cook up an omniscient-I-know-better looking argument to the effect, that it is a common good not to have a place to escape to in summer: Seeing the world while driving the RV is a much more gratifying and enriching experience than vacating in a summer house; or, staying put and making it with what you got is better for the environment and makes for a more profound time to relax and reflect on all the things you don't need of matter, while being rich in spirit. I'm also pretty sure that if I had bought a summer residence, it would give me great pleasure knowing that last year's investment in the belvedere by the lake, already has increased in property value, and I stand to make a bundle if I decided to sell. Everybody should have such a chance, at least people like me – so what about me – in this new context is flipped on its head from the previous common good in that context: From not having to now having a summer residence with two gazebos and a fountain pond full of coy carps, by the lake.

Whatever the social benefit for the group is, it is deemed so according to the whataboutmeist wanting the benefit of cooperation but not chipping in when it comes to the execution of what apparently serves everybody the best. Since you always claim the right to serve asking what about me, the common good installed would certainly stand to benefit you, even if it does not benefit everybody in the group or setting. So, it might benefit others but doesn't really have to as long. It is a smoke screen for a social benefit for everybody to enjoy or an alternate social caring only guaranteed to benefit one. Whataboutmeism has an air of gaslighting to it where projecting expectations by one person may seem as if they are aspirational and socially beneficial to all, but really, they are only guaranteed to benefit the perverted egoist when realized by everybody else also taking the moral fall too if things go wrong. It is a moral position of magic in broad daylight.

Chapter 5
Seeing it coming a mile away

Spotting whataboutmeism may sometimes be tricky. It is often dressed up in what's apparently for the common good according to the one asking what about me while presumably and pompously knowing better. But there are signs of it to watch out for in the process of projecting expectations for others to meet in the aim of gaining control of the outcomes and the situation at large. Once aware of the red flags and gaslighting you can see the whataboutmeist coming a mile away – our own specimen as well as the entire what-about-me-group and hang-arounds of reactionary reproduction.

5.1 Red flags

If SoMe saw whataboutmeism coming a mile away and still makes a bundle on it, why shouldn't we be able to see it too? There are warning signs to be aware of: Does an exchange between Addison and Madison along the following lines sound familiar:

– **Addison**: "You know, I have a cold. I'm walking down to the pharmacy to pick up some cough medicine. See you!"

> – **Madison**: "Well, since you are out, could you go to the lumber store down the block and pick me up three two-by-fours for the new kitchen. If they don't have it there, try the outlet near the interstate highway. Thank you!"

Of course, Addison, had no intention of going anywhere near the lumber store in the first place being sick and all. But Madison now projects the expectation to be realized that while anyway going out for cough medicine might as well be combined with Addison picking up three lengths of wood with a rectangular cross section nominally two inches by four inches. If such lengths of wood are sold out one place, then the gig doesn't end there, because two-by-fours might be purchasable somewhere else albeit presumably further away now the interstate is part of the directions to follow. Since Addison is out anyhow, then go there instead and realize Madison's expectation.

By the projection the mission parameters have changed: Mission is – in line with Madison – now accomplished if Addison secures the exact lengths of wood and thoroughly independently of whether Addison obtains the pharmaceuticals needed to fight the cold. If the mission now fails, it is on Addison. One way it could fail would be if Addison only returns with the cough medicine, but no wood, yet successful if returning with three two-by-fours but short of medicine. It stands to

5.1 Red flags

reason, that the drugs are for Addison's well-being and no suffering from a sore throat, the lengths of timber for a kitchen – but it doesn't have to be theirs, nor have anything to do with their common good.

While the cough medicine is to ensure Lucas' well-being, it is the three two-by-fours for the kitchen that must be secured for the mission to be considered successful … according to Alpha. Strictly speaking, this is not what Alpha says, but rather something that seems to follow from what is being voiced. It is an implicature – something that a speaker suggests or that is implied in what is said without being literally uttered. Such an exchange may very be a red flag for whataboutmeism in the works. It may sometimes give itself away by certain questions or expressions used to project expectations by the practitioner of the day. Some idioms are immediate to see, some more dressed up or downright in disguise:

- "Can't you see that …?"
- "Don't you agree with me that …?"
- "Are you with me in that …?"
- "Other people think/do, so why don't you …?"
- "Didn't you then tell them that …?"
- "I expect that you …"
- "I wish you wouldn't …"
- "I would really wish that you …"
- "I can't understand that you …"
- "I'll be disappointed if you …"

- "I think you should …"
- "If I were you, I would …"
- …
- "I think it is now on you to think of more" … otherwise, I will be disappointed as author of this book. I know what is best for me … and apparently also for you right here and now.

A speech act is an utterance for which the speaker's intention and the possible effect it may have on a listener are the defining features. The speaker hopes to kindle some action on behalf of the audience – from inaction to confusion, from mobilization to purpose. As such, speech acts could be any number of declarations including apologies, greetings, promises, requests and warnings. Speech acts serve their purpose when they are uttered or communicated, as when it is proclaimed at baptism "I baptize you Addison" or "Addison, I apologize for my behavior", in which the apology lies precisely in saying sorry. If Madison were to say: "I expect you to pick up the laths", it implies that a trip to the timber trade is being encouraged, but also tacitly implies that consequences must be expected, if Addison comes back empty-handed, i.e. without three two-by-fours.

Common to the list of questions and expressions above is that they are speech acts as they are intended to

- mislead the target creating **uncertainty**,
- possibly create **false narratives** as to covertly
- making the interlocutor(s) **question their perceptions, decisions and actions**.

It may even get to the point where the target(s) – partners, family members, close co-workers, friends even foes – feel unsure about their perceptions of the situation wondering whether they lost sight of something too important to miss or simply just lost their minds. That's the strategy of gaslighting.

5.2 Getting gaslighted

Gaslighting is a nasty form of emotional abuse and psychological manipulation of a person typically over an extended period eventually causing the victim of gaslighting to question the integrity or coherence of their own thoughts and memories and perception of reality. It may lead to increasing levels of confusion and anxiety, possible loss of self-esteem and confidence, uncertainty as to one's emotional or mental states and stability often enough paired up with an exceeding dependency on the perpetrator. Worst-case abusers are often enough habitual or even pathological liars never backing down when their dishonesty is called out or proof of their deception brought forth. If not downright lies, bullshit, feigned stories or fake news may do the trick on me of distorting the truth and instill an alternate reality in which I will start second-guessing myself.

Rumors and gossip advanced by the perpetrator about you and others around you and the way they apparently look at you according to the "gas-lighter", is intended to discredit, distract and even minimize your thoughts and feelings as misguided overreactions, you are being overly sensitive, even crazy. Besides systematically denying any wrongdoing to avoid taking on any responsibility for whatever the abuser is set to accomplish, blame-shifting is yet another common gaslighting tactic.

If Addison and Madison were to have a profound discussion about why Addison only came back with the cough medicine but no two-by-fours, chances are that the argument soon enough would be twisted such Addison is to blame for the mission failure. Even if Addison was to attempt discussing the emotional impact Madison's behavior has, the conversation stands to get twisted yet again in such that maybe Addison actually is the cause of, and is to blame, for Madison's bad behavior. After making up but revisiting the incident later, Madison-the-gaslight may rewrite history in a way reflecting favorably on the perpetrator but badly on Addison the real victim causing yet more confusion and second-guessing on Addison's part. That's exactly Madison's intention while gaslighting Addison.

The term gaslighting stems from a play by Patrick Hamilton called *Angel Street* from 1938 and was later

5.2 Getting gaslighted

developed into the much-acclaimed suspense film *Gas Light* by George Cukor from 1944. In this movie noir a manipulative husband tries to get her wife to believe that she is going crazy. The bag of tricks to install this conviction includes making subtle changes to her environment by for instance slowly but steadily dimming the flame of the gas lamp yet claiming the room is fully alit.

Of course, Addison and Madison are nowhere near the means put to work by the husband to gaslighting the wife in Cukor's thriller:

- Addison and Madison might not be married,
- they may not even be husband and wife, male or female or something third, *nothing has been mentioned about their respective gender throughout this book*. **They are just humans like the rest of us**.

Whataboutmeism is about power and control no matter sex and possible stereotypes. The gaslighting in whataboutmeism has to do with convincing the environment around you that your projected expectations are aspirational and socially beneficial to all and everybody in the circle, context or situation. If things go haywire and the expectations cannot be realized, the (moral) blame is shifted to the other, never you, given the secular version of the theodicy rehearsed in the previous chapter. And here gender, race, ethnicity, social class, religion and other axes of identity politics are immaterial, even when whataboutmeism is about power and control. The point of departure in whataboutmeism is not about some group affiliation or lack thereof, but rather a certain subject-driven approach to the world.

When you time and time again in assorted contexts or situations put forth a speech act like "I expect that you …", "If I were you, I would …" or questions like "Can't you see that …?", "Other people think, so why don't you …?" etc. they may be signs of gaslighting. If it means that I eventually start to

- doubt my feelings and reality;
- question my judgement and perceptions;
- feel vulnerable, downright insecure or just have to walk on eggshells a lot of the time;
- wonder if I really am what you say I am;
- feel disappointed in myself and what I apparently have become;
- worry that I'm just too sensitive or in some other way inadequate;
- spend much time apologizing, second-guessing myself as to what is right and wrong in our relationship
- and generally have a sense of impending doom if I cannot meet your projected expectations

chances are that you may be gaslighting me as to what the common good of our relationship really is and whataboutme game is on the horizon. Similar, but possibly less bombastic, pronounced or emotionally strenuous warning signs may be found while being gaslighted by other family relations, by friends, colleagues and other parties in your immediate environment.

5.3 The imitation game

In whataboutmeism its practitioner covets what others apparently possess or want. More time with our partner spending time elsewhere, the same profits as friends presumably have from selling their stock, the same success in trout fishing as Håkon the proclaimed victorious angler, the same visibility, recognition, status online as influencers and reality stars, promoting baby oils, sponsored food recipes or the latest G.I. Joe with the Kung Fu grip, seemingly enjoying the same clothes, cars, jewelry, VIP-invitations and beauty products presumably in vogue etc. Why this imitation and why is what is coveted limited to what is already out there in abundance; why not something new and completely different?

Simultaneously with economic growth and advancing individualization, where individuals have devoted themselves to personal development opportunities in terms of recognition, identity, talent and potential and have had the necessary means for this 'self-realization', a convergence still occurrs towards the same trend-things: Four-wheel-drives for urban transportation, expensive watches, high-end apparel, brand labels, designer kitchens and other accessories.

How can it be that everyone is an original individualist and yet everybody converge on exactly the same things? It's neither original nor independent. The basic thesis is that when you don't possess sufficient information to solve a given problem, or if you just don't want to, or have the time for processing it, then it can be rational to imitate others by way of social proof.

Take the market for holiday crime novels. For those of us who are unable to examine the entire market, imitating others' choices solves the problem. We consult the bestseller list, the news' reviews as well as family and friends. This can be rather rational, because through imitation you benefit from the information, which others have gained through, sometimes hard, experience depending on the quality of the crime plot. However, it's not a question of a blind imitation process. Imitation is motivated by the problem that needs solving, and one seeks to imitate those who have had success in doing so. But which problem, except for social status, does fashion solve, and whom should you imitate when everybody are trying to do the same?

5.3 The imitation game

Here publicly private spending has assumed the role that titles and honors had in days of old. It signals acclaim, status and power. These types of signals are naturally social denominations, the meaning of which is only credibly signaled *when each understands that everyone else understands them likewise.* It's like with currency again: We only accept the currency we expect others want to receive. Social status, or "likes" for that matter, is currency just like that.

But when the status signal is no longer embedded in a real value, it becomes a question of a self-feeding process where the individual no longer commands any other quality assurance than the status itself. Unlike money, status or power isn't spent but is simply strengthened when used. *The ones with a reputation therefore end up having a reputation simply because they have a reputation, and celebrities become celebrities for being celebrities.* Just think of the alarmingly high number of reality-show stars produced lately whose only qualification and apparently celebrity impact is that they are like everyone else, and if nothing else, then themselves. But if they can become famous for specifically being themselves, then the rest of us can as well, which is why in principle anyone can become famous for being famous. Status economics may create status bubbles and I could/would/should be part of it too because if for no other reason … what about me?

By the same process we find the answer as to why fashion clothing, fancy cars and the rest of the 'superficial' lifestyle products are perfect elements of status economics, as well as why the consumer bonanza celebration of the individual paradoxically arrives in a standard package. *Imitation is by its very nature limited to the observable*. In a status or power economy, where there are no other guarantees than the social status or power itself, successful imitation is thereby the kind of imitation *everyone can see and everyone can understand* — thus the standard packaging.

Meanwhile, honors and titles only have value if an audience, placed in an asymmetrical relation to the bearer of the social symbols, recognizes them. This is exactly why it's so important that we need to hear about VIP-parties in the tabloids, which the audience is, by definition, barred from participating in. For just as the nobility had to acknowledge back in the day, if everyone gains access to the symbols of power, inflation and then worthlessness follow. While everyone converges toward the symbols of power in order to offset the asymmetrical relation, the content of the symbol is constantly changing. This very feature entails a cat-and-mouse chase in the hope of being initiated in the symbol's latest content. The imitation by definition therefore causes inflation of the symbols of power. But where it implies more knowledge as a by-product in a true knowledge society, it just brings along further spending and demands of renewed capital in order to do so at the crossroad of the information and the consumer society. And voila, that's how it all resulted in a jejune economic crisis and very well may again before long.

5.3 Reactionary reproduction

Whataboutmeism fits this template of imitation perfectly. In the thriller movie, *Silence of the Lambs*, somewhat more brutal and horrific than Cukor's *Gas Light* suspense, Dr. Hannibal "the Cannibal" Lecter lectures FBI agent Clarice Starling about the nature of wanting:

> We begin by coveting what we see every day. Don't you feel eyes moving over your body, Clarice? And don't your eyes seek out the things you want?[11]

"What about me?" implies wanting something others already have or desire and that's exactly "coveting what we see every day" precisely confined to the observable which imitation by nature is limited to. In terms of the common good — besides being arbitrarily fixed by the whataboutmeist in question and subject to change at will — it is always defined within the framework of the already established which exactly what you want a piece of. A common good is hard to establish if nobody apparently cares about it, whatever it is. That would leave whataboutmeism without its fundamental relational component. So, proclaiming a common good that others presumably care about is essential to whataboutmeism which lights the match of the imitation game inherent in what about me. Relying on a common good that others seem to value is essential to whataboutmeism. And what others seem to appreciate is the essential component that in turn stimulates the imitation game.

The good thing about the imitation game is that it makes its players predictable – you can see them coming a mile away. Yet the bad thing is that the price of successful imitation is high. There is no divergent, untraditional or creative thinking neither present nor required for whataboutmeism. Same old s***, different day. It is in plain sight reactionary reproduction of what is already there and yet more of the same to fuel forever more and even stronger demands of having it my way. If someone is thoroughly unimaginative, uncreative and uninspiring but insistent on getting it their way that's likewise a red flag for whataboutmeism to be aware of. No wonder now that some

- **bullet items**,
- **numbered items**,
- **list items**, and
- **text boxes**

5.3 Reactionary reproduction

throughout this book are emphasized in color **red** as they are red flags characterizing whataboutmeism. If only interested in the executive summary, read the stuff in red:

> **Whataboutmeism in so many words**
> *Take everybody out trick-or-treating to get the candy, retain it all yourself, gaslight and bamboozle your mates and hang-arounds into the belief that their treat was to help you get all the M&Ms, Reese's Peanut Butter Cups, whatever and Whatchamacallit. Then do the same next year, and the year after that, and … . As long as you feel triggered, entitled but unenlightened, knock yourself out.*

Chapter 6
Separate spreadsheets

Whataboutmeism comes with triggers typically resulting in some emotional response. Feelings of anger, rage, jealousy or disappointment are indubitable as long as the person in question entertains the emotions in question. The indubitability itself sparks an entitlement to have it exactly my way such that the anger, rage, jealousy or disappointment subsides. It's on the other hand for external surroundings or the opposing terrestrial to realize the receding emotional tide in the subject. And that may indeed be exceedingly difficult for external parties if all the expectations and priorities are indiscriminately pooled into just ONE zero-sum spreadsheet for which the whataboutmeist is the creator as well as the accountant.

6.1 What *you* do, triggers *me*!
A good friend of mine reported from an exchange at a work-related meeting. On the table among the colleagues that day was the implementation of a new, and to some of the parties present, rather controversial business strategy. As deliberations went on, so did more heated discussions albeit still civil. During a break, a

party to the conversation, pulled my friend aside and announced:

- "I'm really triggered by what you said, and now I'm angry",

to which my friend replied:

- "Well, I'm not – so being angry is your problem, not mine."

Being triggered is to elicit a strong negative emotional reaction of shock, fear, worry, jealousy, disappointment or anger often connected to previous unpleasant memories, past bad experiences or even traumas. It may be voluntary or involuntary, often enough very powerful and it may not always be immediately clear what sort of thing will set it off – neither to the party in question, nor especially to the social surroundings.

Although my friend's reply may come across as arrogant, it doesn't have to be. You can't ask of my friend to be able to necessarily anticipate what would trigger the other party's anger especially when the conversation is kept civil even while (strongly) disagreeing. Triggering anger may accordingly be a disproportionate response on my part if no such thing was intended on your part. Of course, I could also reply:

6.1 What you do, triggers *me*!

- "Well, the fact you are now angry really triggers me, so now I'm at least as angry as you are, but possibly even more."

That may initiate a war of attrition as a function of our ever-elevating levels of anger towards each other most likely getting us nowhere. A third possibility would be for my friend to say in response:

- "Nah, you don't really feel angry, you just think you do."

That is likely not a winning strategy and would indeed anger the disagreeing party beyond belief. My emotions are reserved exactly for me and are indubitable too as long as I have them, so who are you questioning my emotional state? Thus, on one hand, arguing on emotions is an extremely powerful strategy as your opponent cannot second-guess your feelings nor challenge them with a counterargument. But at the same time, it is a very weak strategy as further reasoning together, communication and cooperation stop right there in their tracks. When a quarrel is looming at large, Madison tells Addison: "I'm not in the mood to talk about it!", then that's it! It may very well be quite important to talk about, but not being in the mood apparently slams the door shut with Madison and Addison in each of their (mental) confinements.

6.2 Emotional tyranny

At the 2022 Academy Awards show something unusual happened never seen before in the Academy's history of almost 100 years. Actor Will Smith slapped Oscar's host, actor and comedian Chris Rock, across the face for a joke Rock made about Jada Pinkett Smith, the wife of Will Smith. Chris Rock had cracked a joke which the Smiths thought was as insensitive as it was indecent, injuring the family across the board. A giant debacle came of it afterwards. Hardly anybody remember who won best picture, best actress or actor in a leading role, best sound, but everybody remembers the violent incident broadcasted around the world for everybody to watch live. It was better than going to the movies.

Will Smith's initial attempt of an excuse was that his feelings had gotten the best of him. Engrossed in your emotions is not a viable excuse for physically hurting a fellow human being if not legally a crime of passion – which, considering the circumstances, probably is a bit farfetched, even between Will Smith and Chris Rock. Neither would it have been a viable excuse asking Chris Rock to leave the stage just because Will Smith and co. felt insulted. Emotions are not powers of veto.

It is well within one's right to express sentiments of being hurt or injured by what someone decides to say. One may even go as far as to say that this someone voicing an insult is obligated to listen to how you feel

6.2 Emotional tyranny

about it. But exercising physical, mental or social violence by expulsion or stigmatization is not the answer. It's *your* emotions, and thus the individual that is faced with the problem of staying or leaving and not resorting to violence. Our individual emotions are not to be considered the borders of collective freedom. If they were, individual emotions of disappointment, jealousy or anger, may turn into instruments of power to be exercised over parties I don't like, who stand in my way or somehow stand for something I find repugnant or unpleasant. Not mindset policing but emotional policing, which taken to a group level is called affective polarization. "It's not necessarily your opinion I strongly disagree with, I just don't like you" would be the idea of affective polarization between groups. On this basis you may start banishing or cancelling people you meet in the local supermarket, at college, on the bus or at work if their mere appearance, political opinion or what have you affect you emotionally. You may try to sanction your partner too while at it.

Former US Ambassador to the United Nations and sociologist, Daniel P. Moynihan (1927-2003), once wrote in a column for *The Washington Post*:

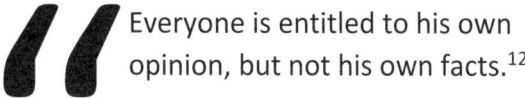

Everyone is entitled to his own opinion, but not his own facts.[12]

In the same vein, you are entitled to your own emotions, but you are not entitled to let them rule others just because you feel what you feel it – or feel like it. Being entitled to your feelings is very different from being entitled to have it your way just because your emotions are indubitable to you. The indubitability of emotions make for the strongest and weakest arguments in debate at one and the same time. Who am I to question or second-guess your emotions when you have them? That's the strong part. The weak part is that our exchange is likely to halt right there and not move an inch ever: "This is how I feel period!"

It is often thought easy to contract on an opinion when in agreement. Curiously enough a contract of cooperation is also required if the parties in question commit each other to *enlightening disagreement*. If disagreement just boils down to "now we leave it there" then we are left with unproductive incommensurability of "my opinion is this; your opinion is that – oh, and by the way, you're a moron as you either can't or won't understand my thoughts and feelings". Incommensurability leaves us to our own oddities making difference the centerpiece where commonality should take center stage. The salubrious outcome of recognizing commonality is empathy, understanding and fundamentally seeing the other as an individual which could have been any one of us under similar circumstances.

6.2 Emotional tyranny

Hence when in disagreement an indispensable task is to understand why the other person believes whatever is believed about politics, relationships, children, family, friends, co-workers etc. The interesting thing is not whether we agree or disagree, the interesting thing is not, that I believe P while you believe not-P. The important thing is *why* you are of one opinion yet I of the other. That may turn into a lot of work figuring that out in detail. First, I must sit back and listen to your arguments, dissect premises from conclusions, keep things properly apart, bring hidden or tacit, possibly controversial, assumptions to light and eventually consider whether I deem your reasoning valid or not. And then, I expect you to do the same for me. Then either we'll continue the discussion or end up in agreeing to disagree.

This process is enlightening either way but requires extensive cooperation between you and I. The worst thing that will come of it is to become the wiser. Not only do I now know that you are of such-and-such opinion, but I also know why! *The enlightening disagreement is the cooperative disagreement.* This presupposes that we can sort things properly, not try to pool everything indiscriminately together into one spreadsheet for everybody else to measure up to. Some things are apples, others are oranges – true both have spheric shapes but that's about it for comparison. Some problems relate to the partner, others are about the children, some issues are associated to friends, or foes for that matter. The three items cannot be ranked in the same spreadsheet. They are essentially different issues.

6.3 Keeping separate spreadsheets

Suppose the imaginary tale of Addison and Madison is the real-life example of two parents opposite gender stepfamily constellation with kids from previous encounters described among the scenarios of whataboutmeism in chapter 1. Madison has the nuclear background with unbending one-model-fits-all-principles, Addison's parents broke up early, found new partners with yet their children and new workable constellations were formed from patchwork baggage using pragmatics schematics. Assume additionally, for sake of argument that Madison views life, not by approximation, but a decidedly zero-sum game. Everything that Madison doesn't get goes one-to-one to others, in context of a couple, that would be Addison.

To ensure equality, fairness and transparency among the new members and provide the best conditions for sustainable best stepfamily dynamics, Madison sets up ONE universal spreadsheet to keep track of things. It consists of a numbered and accordingly ordered positive list of priorities ($a_1, a_2, a_3, \ldots a_n$) based on a utility score of Madison's making. Having good report with Addison has top priority and is awarded a utility score of, say,

6.3 Keeping separate spreadsheets

100. And so, it goes for the remaining finite number of priorities up to and including priority a_n and its associated utility score X_n.

Interesting side note: Although this may seem a farfetched way of running a stepfamily, I once had a stepfather who reportedly set up a similar spreadsheet adding up the pro's and con's when he was to decide whether or not to move in with my mother. In the end, the sum of positives outweighed the final sum of negatives and thus it came to pass. All for the better, so Madison's mode of operation is not completely unprecedented:

		Priority	**Utility**
1	a_1	Good report with Addison	100
2	a_2	Good report with kids	90
3	a_3	Good report with family	79
4	a_4	Good report with friends	65
5	a_5	Good report at work	59
⋮	⋮		⋮
n	a_n	Good report with …	X_n

Read the listing in the following way: As humans, we have priorities, and their importance is arranged according to how much they mean to us. Put differently, how much utility is obtained from fulfilling the priorities in question.

Now, say that the utility (U) of priority (a_i) realized in Madison's (M) spreadsheet is some number X_i associated with a_i, so $U_M(a_i) = X_i$. Madison's design of life as a zero-sum game bids that whatever Madison doesn't get befalls somebody else, namely Addison. The proportionate disutility $-X_i$ of the same prioritized expectation not realized for Madison consequently becomes Addison's gain, written $U_A(a_i) = -X_i$. What Madison loses out on, denoted $\hat{a}_i$, is exactly Addison's gain, a_i, which is why

$$U_M(\hat{a}_i) = -U_A(a_i) = -X_i.$$

Loss and gain are opposites in a zero-sum game between Madison and Addison. If Addison values a_i and gets it consequently means Addison doesn't get it. The overall calculation related to utility becomes

$$U_M(a_i) + U_A(\hat{a}_i) = 0.$$

Zero is the answer as prescribed in a zero-sum game. Madison's priorities and utility assignment spreadsheet is an ordered set arranging all entries in the spreadsheet according to Madison's assigned priorities based on utility.

As may be read off Madison's spreadsheet, a good relationship with Addison has a utility value of 100, while

6.3 Keeping separate spreadsheets

good relations at work, by comparison, only have utility 59. So, a good relationship with Addison has higher priority and is thus worth pursuing more than say good report at work. This ranking applies to all entries in the spreadsheet depending on the utility values assigned. The utility value assignment thus gives rise to an ordering between all items which reflects aspirational worthiness according to Madison. It applies to everything found in the spreadsheet and the ordered set of priorities. An isomorphism is a mapping that preserves structure from one set to another and is reversible. This is exactly what Madison's design of life as a zero-sum game over the spreadsheet does. Life as a zero-sum game in the spreadsheet is just a function that takes the utility value realization of an arbitrary priority a_i of Madison and maps it into Addison's loss, $â_i$, and vice versa, when Addison's gain becomes Madison's loss. The function does so while preserving the order of priorities in one-to-one correspondence. And that makes the feature consistent with Madison's inverted and weird approach to life when the zero-sum game rules:

> **Everything Madison doesn't cash in on in terms of priorities or expectations is exactly what Addison is rewarded. It doesn't matter whether Addison wants that at all. Addison may in principle have completely different priorities and utility value assignments. But that wouldn't even**

> **matter on Madison's view: Either you get what you want, or you don't and if you don't ... Addison get's it and again, Addison's gain is Madison's loss. End of story!**

Addison is not just Addison but a proxy for any other party Madison is up against in a life designed as the zero-sum game — among friends, at work, online — all for which similar balance sheets may be set up. Nevertheless, such unfortunate isomorphic mappings between gains and losses of Madison and the social surroundings are exactly what you get out of ordering all priorities in the same sheet without distinction and then balance a zero-sum game on top it.

But ONE balance sheet for everything is setting yourself up for failure and whataboutmeism. As humans we prioritize constantly, keep track of what came to pass, and what didn't, and balance out pros and cons of actions navigating through the stream of life and those in it. What is important however is maintain separate spreadsheets for essentially different priorities and not throw apples in a bag full of oranges. Addison spending time with a child from a previous marriage belongs to a set of priorities having to do with Addison, and not Madison — and vice versa. The two sets of priorities are in this regard *incommensurable* and cannot be ordered in accordance with an order isomorphism.

6.3 Keeping separate spreadsheets

Similarly, a work promotion for one party, does not entail a loss for the other – they are essentially different things. They may indeed benefit both parties in the same way that Addison spending time with an offspring of previous encounters, without Madison insisting on and commiserating about the ONE spreadsheet to rule them all, may have a derived beneficial effect on Madison and Addison's relationship at large. Addison is no longer between a rock and a hard place, and Madison doesn't have to keep the scoreboard on this entry as they are simply on two different spreadsheets. Such a separation between spreadsheets and the entries in them will set both parties free because it is simply not all about you. Sometimes it is on somebody else's spreadsheet but then you don't have to do the book-keeping. And honestly, how much fun and forward-looking is it really to be lead accountant in a romantic relation?

Separating spreadsheets with essentially different entries is also advantageous in family affairs at large, in relation to friends – high maintenance friends especially, colleagues, during recreational stock-trading and short-selling comparisons, while comparing yourself to other users online and when you get an RC car for your birthday that's not for you – but your friends – to try out on the day of celebration. Me not trying it out, is not about me, but about my friends racing around with it, and that has little to do with me missing out, but much to do with them having fun. I may even cherish the fun they are having as a free and gratifying bonus, without feeling deprived of anything but rather blessed with something – their fun. It's not about me, but about them.

Don't forget to run different spreadsheets for things essentially different was the point my father was trying to convey to me back then. It'll save you a lot of self-inflicted pain and grief when you don't have to keep score and constantly compare yourself to the gains and losses of others that substantially has very little to do with you if anything at all. Get it straight! Maintain separate spreadsheets for essentially different things. That'll set yourself free and the ones around you. If you don't, whataboutmeism is right there for the taking making you possibly either a coward or a crook.

Chapter 7
Whataboutmeist: Coward or Crook?

How much of whataboutmeism is really about cowardice behavior and risk aversion, and how much is about being a black hat and taking risk? It is as if you want the benefit of cooperation and community but refuse to chip in, wanting something for free for others to realize. A coward lacks the courage to face danger or take a chance, but still wants opportunity or advantage. A crook is dishonest, conniving even manipulative, willing to chance it prospecting a good return. When you practice whataboutmeism, are you a coward or a crook? Now, that's the question.

7.1 Forget trust – it's overrated

Whataboutmeism is about control through projections of expectations for others to realize a common good in no small portion hoodwinked by the morally exonerated practitioner. Since the whataboutmeist apparently knows better what the common good is, given the secular theodicy argument, the protagonist simply doesn't *trust* others to decide what is best for the bunch or the world at large. It's about a lack of trust which breeds uncertainty on behalf of the flagbearer and this precariousness is not conducive for control. The tricky part is that trust and uncertainty is a package deal.

Trusting somebody means essentially that

1. we are vulnerable to, or uncertain about, others,
2. we, by and large, think well of other people at least in situations where it matters,
3. we believe others to be competent in what they do at least situations that concern us.

If Addison comes to Madison with a delicate issue concerning, say, some intimate constraints on their sex-life, it comes with a conviction on Addison's part, that Madison will refrain from posting about it on TikTok, Instagram or some other social platform for the world to see. After all, it is between the two of them, their business only. A delicate issue to be handled with discretion by both parties involved. Should Madison nevertheless decide, for some reason or the other, to make a video about it and post it on TikTok, that certainly would be considered a breach of trust by Addison.

Trust presupposes the possibility of a breach which induces a vulnerability or uncertainty about others as relayed in item 1 above. American writer Terry Goodkind (1948-2020) explains this in a clear, crisp and concise way when he relayed: "Only those you can trust betray you". If you don't like somebody and find them

7.1 Forget trust – it's overrated

generally unpleasant, why would you entrust them with anything at all, item 2 above; and if they prove incompetent in areas of concern, discount them too as somebody not trustworthy, item 3 above.

Now, distrust is not merely the absence of trust. You may neither trust nor distrust someone thus being suspended between the two states. Due to the possible suspension, trust and distrust are not exhaustive, although exclusive. On the same matter it is impossible to both trust and distrust someone at the same time. Distrust is not just non-reliance either. At work, Addison may choose not to rely on the assistance of a colleague, not because of distrust, just because the colleague is overly busy. But whereas non-reliance comes without a normative dimension, distrust does.

Suppose Addison distrusts Madison for no good reason, and Madison finds out about it. Chances are that Madison would be upset, angry or hurt to learn about Addison's lack of trust in the partner. Conversely, Madison will not necessarily be accompanied with the same reaction being told that Addison didn't choose to rely on Madison for some reason. While distrust is a bad thing, non-reliance doesn't have to be in any way, shape or form. So, distrust comes with certain action-tendencies of avoidance or withdrawal making distrust incompatible with reliance, the same way that trust is compatible with reliance. Addison can be forced to rely on Madison although distrusted. Even then, Addison will attempt to keep Madison at safest possible distance.[13]

Trust and whataboutmeism are like oil and water – they don't mix well:

1. A whataboutmeist doesn't want to be vulnerable – due to the possible breach of trust – as it doesn't rhyme with control as little as uncertainty does.
2. Given the "formula of whataboutmeism", independently of whether you think well of people around you or not, they are primarily means to realize your ends as you act in keeping with the lopsided Kantian imperative that you act in accordance with the maxim, that others are to cash in your expectations.
3. If you thought others around you were competent in matters of importance, like defining the common good, you might outsource this ordeal, but you don't, and you have the secular theodicy argument to back you up.

Signing the bill

- that trust, and distrust are not exhaustive, but exclusive,

- that distrust comes with a normative dimension which non-reliance doesn't,
- that distrust is incompatible with reliance, while trust is,
- distrust animates to avoidance and withdrawal whereas trust doesn't

is all uncontroversial to the whataboutmeist. So much more the reason not to pursue trust and its derivatives any further. Alas, trust may be good, but control is better still. Bottom line, after going through the checklist, trust is not something you rely on as a whataboutmeist: Either because you are a risk-averse coward or a conniving crook complete with illusions of grandeur of consistently knowing better. To settle the score between coward or crook, how we feel about winning or losing and taking chances is to be ironed out first.

7.2 Don't chance it!

To answer whether a whataboutmeist is a coward or a crook is a complex one with quite a few moving parts to it between the psychology of choice, risk averse versus risk seeking behavior and duly distributed probabilities. The point of departure is psychology professor Daniel Kahneman's, by now, legendary studies on human decision making and its processes. Kahneman unpacks these mechanisms for a broader audience in his 2011 bestseller *Thinking Fast and Slow*. A book that may be viewed as a reader-friendly mainstream testament to Kahneman's impressive contribution to the theoretical understanding of decision making and the psychology of choice. Kahneman's studies started in the 1970s while collaborating with distinguished cognitive scientist and mathematician Amos Tversky (1937–1996); the two came to develop what they dubbed *prospect theory*, for which they were awarded the Nobel prize in economics in 2002.

Ever since its inception, prospect theory has been applied in a wide variety of economic studies from consumer choice to labor market demands and insurance policy making. It is nowadays considered an integral part of behavioral economics. Kahneman and Tversky came up with a behavioral model that maps out how humans choose between different alternatives that involve risk as well as uncertainty. An important key point is that we as humans are *risk averse*, meaning that we avoid exposing ourselves to risk if possible: We care more about what we may gain, than what we equally might lose. If there is a high probability of a smaller gain, then rather win little with certainty by being risk averse, than be a *risk seeker* with higher but less probable gains.

Think for instance of behavior online as it relates to posting material on a social platform: Suppose there is a high probability of receiving a lot of likes on your next post, because your opinions align perfectly with your

7.2 Don't chance it!

circle of friends, you will ultimately only fear being terribly disappointed and as a consequence avoid drawing your opinions more sharply —in this situation you've become *risk averse*. Conversely, if you assess that there is a high probability that your circle of friends does not fancy your opinion of, a former president, but you nonetheless feel obligated to ventilate your opinion, you can merely hope for no further loss of standing when you take the chance and post it as it is – in this scenario you're more *risk seeking*.

Suppose your tell-it-all-video of a celebrity is of such poor quality that one hardly can tell what is going on, but is still good enough to potentially harvest a lot of likes (if lucky). Then you might as well send it into circulation without seeking permission to do so from the one you're about to expose. Risk seeking behavior again. And finally: You become risk averse, if the current version of the TikTok video you're contemplating uploading has a slim chance of failing, can generate a lot of likes without having to add extra steps and a 'cheeky move' in fear of losing big.

The gains, losses and willingness to take risks according to Kahneman and Tversky depending on the probabilities of winning or losing, say, $10.000, may be summarized in the 2x2 matrix below. The top row includes the scenarios likely to happen, that is, gains or losses with high probability, whereas the bottom row comprise the scenarios unlikely to happen, either gains or losses with low probability. In each cell there is an outcome, for instance, "95% chance to win $10.000", a corresponding emotion evoked, "Fear of disappointment" and the type of behavior associated it, say, "RISK AVERSION". Finally, the certainty effect signifies the general tendency for individuals to prefer certain outcomes rather than probable ones. The possibility effect refers to situations in which individuals tend to overestimate the probability of an outcome just because it is possible. [14]

	Gains	**Losses**
High probability	95% chance of winning $10.000	95% chance of losing $10.000
Certainty effect	Fear of disappointment	Hope to avoid loss
	RISK AVERSION	RISK SEEKING
Low probability	5% chance of winning $10.000	5% chance of losing $10.000
Possibility effect	Hope of big reward	Fear of big loss
	RISK SEEKING	RISK AVERSION

"Gains" and "Losses" may be substituted respectively for "Get my way" and "Compromise" for purposes at hand. Even though $10.000 probably won't buy you the

7.2 Don't chance it! 109

common good you desire, that's what ten grand represents in the matrix.

Now, if there is 95% chance of hoodwinking a common good for others to follow for me to get my way, all I fear is disappointment should it fail. I'll be damned if I'm going to risk anything much by suggesting, say, something even more demanding but of (potentially) greater benefit … at least to me. No way I'll take that chance! I'll be risk averse in the top left scenario. On the other hand, if there is a very slim chance of getting it my way anyway, I might just try to tighten the screw a bit more by possibly suggesting something even more outlandish as a common good for others to follow for my eventual gain. As I stand to lose anyway, all I fear is an even bigger loss, so why not just take the risk seeking chance in the top right cell? Similarly, if there is just a 5% chance of bamboozling my common goal through for everyone to pay homage to, then there remains a slight chance to do so and score big for whataboutme, so I might as well stand up and risk it hoping for some great reward will be coming my way. A case of risk-seeking. And the final scenario of risk aversion in the bottom right cell; fat chance of 5% of not getting my way pace the common good of my choice, not much chance of losing, so no need to risk it as that could really turn into a big loss. Better play it safe.

7.3 A relationship gone south

Two immediate takeaways from the 2x2 matrix and the example above;

(1) the prospect of losing a potential gain, result in feelings of loss aversion, whereas
(2) we are ready to seek risks once faced with near-certain losses.

Gains and losses are valued differently. More weight is placed on perceived gains vs. perceived losses. Hence individuals tend to make decisions based on perceived gains rather than perceived losses. Suppose you stand to get $50. Either you get them outright and immediately at the cashier's booth, or you can get $100, but then again you must return fitty bucks to break even on getting $50. The utility of the $50 is the same either way. Be that as it may, individuals are more likely to opt for the straight cash of $50 up front as this single gain generally seems more favorable than having even more cash, $100, and then suffering the loss of returning $50.

7.3 A relationship gone south

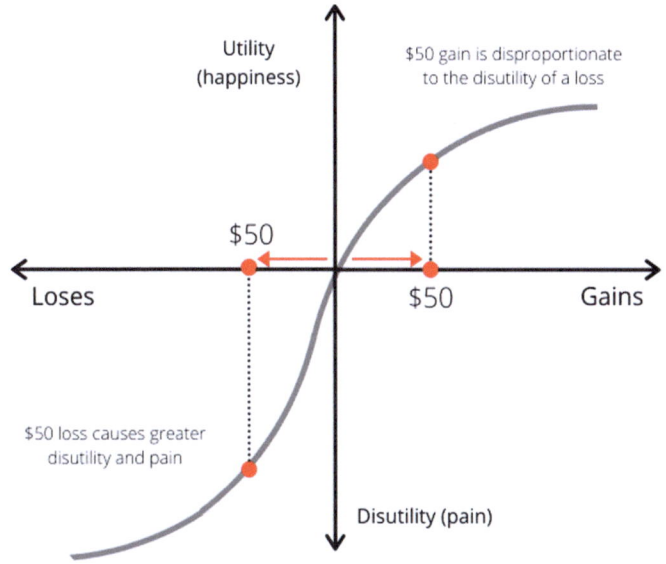

Losses have a greater emotional impact than does the equivalent amount of gain – "losses loom larger than gains" in the notorious words of Kahneman and Tversky.[15] Thus, when choices over the same outcome are presented in two different ways, one emphasizing gain, while the other loss, we tend to go for the option offering perceived gain. Returning to the fifty-dollar bill, losing it feels about twice as bad as gaining it. From here, the larger the loss, the more aggressive the pain, and the equivalent gain, doesn't add to the same inverse level of happiness as Kahneman and Tversky ascertain.

Revisit finally the relationship of Addison and Madison. Who are they – cowards or crooks – considering gains, losses, happiness, pain, risk averse or risk-seeking behavior? As indicated all along, Madison has more riding on "get my way" than Addison seems to have given the payoff distributions fixed in chapter 2, section 2.3. The specific payoffs for the different outcomes are of little concern right now. Four scenarios to consider associated with Addison and Madison's relationship which turn out to correlate risk aversion with cowardice behavior (**1**, **2**) and risk seeking with crook behavior (**a**, **b**):

- **RISK AVERSION and COWARD (1)**. If Madison thinks there 95% chance of "get my way" and still stay in the relationship with Addison, then don't chance it, as that could lead to a break-up. That loss would be disappointing to say the least even if Madison may not be overly enthusiastic about the general state of the relationship considering their track-record playing the whataboutme game. So Madison will be risk averse, and somewhat of a coward too: Rather stay in a rocky relationship and have it your way than face the music.

- **RISK AVERSION and COWARD (2)**. Slim chance of 5% of not getting your way yet stay with Addison, that's a no-brainer for reasons similar to (1). Madison will become risk averse, subject to the possibility effect of overestimating the big loss,

7.3 A relationship gone south

and thus a coward staying put in an unstable relationship with Addison.

- **RISK SEEKING and CROOK (a)**. If there is a high probability of not getting your way, 95%, while not staying in the relationship with Addison, that would be a big loss. Hence Madison might as well take the chance and try tightening the screws on Addison even more by either upping the ambitions of the common good goal or invoke excessive sanctions if the goal currently entertained is not fulfilled by Addison. As Madison stands to take a beating anyway, but there is a possibility, although very improbable to take home the score, then take the chance with the neglectable hope of skirting an even greater loss. That's risk seeking and conniving crook behavior while at it.

- **RISK SEEKING and CROOK (b)**. Finally, 5% chance of getting it your way and still be with Addison is a fat chance. Madison might as well give it a shot, bamboozle Addison again for Madison to possibly take home a big win, although chances are against it as Madison overestimates what is probable given what is possible. Risk seeking behavior again from a corrupted crook.

And now for the curtain call question: How much of whataboutmeism is really about cowardice behavior, and how much is about selfishness or corruption? As we tend to make decisions based on perceived gains rather than perceived losses, then salvaging the relationship while getting your way is much preferred even if you are overall somewhat displeased with how things are going with you and your partner. If such scenarios exist, stay risk averse but a coward. On the other hand, if you stand to lose the battle and the war either way, the hope of avoiding a big loss or hoping for a grand reward against all odds, may turn on risk-seeking and the selfish crook in you. You start tightening the screws on your partner with unreasonable goals to be meet while excessive sanctions are enforced. It is all just set up for you on the slim chance of getting your way while hoodwinking your companion, comrade, chum, consort, colleague, confederate or coconspirator.

On an endnote Kahneman reminds us that **RISK SEEKING and CROOK (a)** is really where the rubber meets the road and worse decisions are made in response to already bad options:

> Many unfortunate human situations unfold in the top right cell. This is where people who face very bad options take desperate gambles, accepting a high probability of

> making things worse in exchange for a small hope of avoiding a large loss. Risk taking of this kind often turns manageable failures into disasters.[16]

Personally, I can sign that bill looking back at some past relationships and one previous marriage. That's the reason why they are in the past rather than the present. Don't be a coward, don't be a crook. To shrug off the dress of whataboutmeism requires you to understand and appreciate the dynamics and benefits of trust and, very importantly, cooperation.

Chapter 8
Work the problem, work together

Not falling into the whataboutmeism trap starts by dropping the idea of viewing life as a zero-sum or something awfully close to it. Life is not to be approached as a game of chicken, nor a prisoner's dilemma. But rather a cooperative game to the end of collectively gaining more than what may be obtained on an individual basis. Cooperation is tricky and requires discipline, but it does pay off in the end for everybody involved – if we *trust* each other.

8.1 Hoarding dog dumps

Addison and Madison have a dog – Bandit. It must be aired ever so often as such creatures do – they must take care of business. This routine task of their relationship relies with Madison. As much as Madison enjoys taking Bandit for the morning walks around the neighborhood while sipping coffee from the mug, there is one chore associated with the otherwise pleasant stroll around the neighborhood that doesn't sit well with Madison: Bandit's defecation coupled with the question of whether Madison should pick it up or leave it to its own devices. If everybody else decides to leave it be, then what about me? Why should I not leave it be? If everybody decides to pick it up, then again, what about me?

Return to game theory for a systematic treatment of the problem. The dog owner's immediate dilemma consists of the following choice: "Pick up" or "Leave it". When this decision is viewed in isolation, any dog owner prefers, like any creature, to perform "Leave it" rather than "Pick up". If this were not the case nature would have created dogs with the ability pick up after themselves. To describe Madison's predicament, construct a model, where the number next to the action represents the subjective utility the action has to the individual dog owner *relative* to the alternative – or, simply, what I get out of the actions I may perform.

Madison		
	Pick up	1
	Leave it	2

Everything else being equal, Madison also prefers to "Leave it", rather than the alternative "Pick up". So far so good, assuming Madison is the only operator and Bandit the only dog around. That's not how it works around the neighborhood. There are other dog owners, and they have their own Buddys, Bellas, Cocos and Dixies that likewise must take care of their morning glory. To be considered is henceforth Madison's valuation of the possible outcomes above when

8.1 Hoarding dog dumps

combined with actions performed by other dog owners. That means expanding the game matrix accordingly:

		Other dog owner	
		Pick up	**Leave it**
Madison	**Pick up**	1, 1	-1, 2
	Leave it	2, -1	0, 0

Here comes the travesty: This new game matrix may be used to show how any dog owner's individual choice, including Madison's, determines an outcome which is as tragic as it is ironic and both at the same time.

Initially realize that the two dog owners prefer the outcome where they both choose to "Pick up" (valued by both parties to have utility 1), compared to the outcome where both choose "Leave it", (valued by both to utility 0). This reflects the fact that dog owners, Madison included, just like all other sensible people, prefer clean sidewalks rather than eventually being neck deep in dog dumps. But how can it then be, as observed in many a township, that sidewalks by and large form an utter museum of droppings from all possible species of dogs at least until the MSCS – Metropolitan Street Cleaning Services comes around?

An answer is forthcoming when taking a closer look at the model to see how dog owners' individual desires and actions interact to produce the tragic outcome. To gain precision first make sure that the model applies to the real world where there are usually more than two dog owners in a given neighborhood. In this situation it still holds that they all share the same predicament. This may be incorporated by assuming any given dog owner to take up the role of Madison, while the role of 'Other dog owner' is set to represent the aggregate decision of the remaining dog owners in the entire neighborhood. Thus assuming that all the other dog owners choose to "Pick up", this corresponds to the generic 'Other dog owner' choosing to pick up. With this generic feature at hand, one may study the choice-predicament faced in this situation by any given dog owner in terms of Madison.

So what does Madison do, assuming that Madison is both sensible and expects everybody else to choose "Pick up"? Madison is now confronted with the decision between "Pick up" and "Leave it" in the situation where all other dog owners have chosen to "Pick up". To study Madison's mare's nest, just focus on the game's first *column* and identify the action, which is related to the best result for Madison. The choice reduces to the first simple game matrix for Madison alone for which any sensible person would prefer to "Leave it". Viewed accordingly the case remains the same since one single dropping doesn't ruin the possibility of being able to walk comfortably down the street after all, and simultaneously one avoids the prickly experience of

8.1 Hoarding dog dumps

picking up the hot stuff. In other words, one dump left behind makes little difference to the overall accounting sheet for excrements. Given that Madison expects everybody else to "Pick up", Madison will prefer to "Leave it". Of course, the problem is that everyone shares this impasse – everyone is Madison from their perspective. Thus, *if it is expected that everyone else chooses to "pick up" after their dog, then each individual dog owner prefers to "leave it"*.

Next, consider the situation where each individual dog owner, that is Madison, expects everyone else to choose "Leave it". Studying Madison's quagmire, it suffices to focus on the second column of the game. Under these circumstances Madison still prefers the action "Leave it" to "Pick up"! After all, the reasoning goes, a single "Pick up" won't save the tragic condition of sidewalks anyway, and favoring accordingly one again avoids the uncomfortable and demanding experience associated with "Pick up". In fact, notice that if all others choose the "Leave it", while you choose the action "Pick up", then you end up with the worst possible outcome, where you must pick up as well as be forced to walk around in all the other dog owners' 'guilty conscience' — therefore the negative value -1 for this outcome.

Finally, revert to the point that each individual dog owner from their perspective is Madison. This ultimately means that the tragic and ironic conclusion follows: No matter what all dog owners are expected to do – "Pick up" or "Leave it" – each individual dog owner prefers to "Leave it". Tragic and ironic since this leads to the outcome of the game, which *all* dog owners find less satisfying than the outcome where *all* choose to "Pick up".

Even though every dog owner agrees to the common goal of tidy sidewalks, parks and public squares, dog owners' individual desires and actions may interact in ways to produce a sub-optimal result to say the least. Of course, it may help if non-dog owners decided to help. But I have yet to see such a human creature of the not-dog-owning kind, scoping around with plastic bags or a scooping up arbitrary dog dumps encountered on the way.

What about me – what do I do depend on what others decide to do – when we all know what's the right thing to do – is a pickle showing up here, there and everywhere. During the first stages of the COVID-19 pandemic as the entire world virtually went into lockdown in the spring of 2020, politicians, lawmakers and government institutions around the planet warned citizens against hoarding commodities like toilet paper, yeast, wheat, canned food and baby formula. This could lead to empty shells in convenient stores and a supply shortage looming at large. Not to mention everybody running to the stores also puts public health in jeopardy

8.1 Hoarding dog dumps

by people standing too close together possibly contracting the virus.

Here is a piece of reasoning supporting hoarding when everybody individually asks themselves what about me?

- If everybody buys normally, it'll be good for me to hoard and secure what I need.
- If everybody hoards, I'll have to hoard, not to wound up by the end of the line where there is nothing left.
- Thus, no matter whether others decide to hoard or not, I'll have to hoard all the same.

If that's the way the individual thinks, that's the way everybody thinks what about me and the supply shortage is well within the realm of possibility. Even when authorities told us to cool it, there was really no incentive not to hoard, the alternative was always worse.

Madison was also faced with this exact quandary during the COVID lockdown, and it came just while walking Bandit and considering the defecation pick-up-problem on the way to the store to buy toilet paper:

		Other people	
		Buy normally	**Hoard**
Madison	**Buy normally**	3, 3	-2, 6
	Hoard	6, -2	1, 1

Madison hoarding while other people buy normally is a payoff 6 for Madison, -2 for others: Two weeks supply of toilet paper for Addison and Madison, and the rest must do like dogs. If everybody buys normally, Madison included whatever you need now, you'll get and there is a 3 payoff for all. Whereas, when Madison and all other parties hoard it is under stressed and potentially unsafe shopping conditions, so subtract two utility units for each and all, 1 for Madison, 1 for any other too.

There is only one Nash equilibrium where all parties' respond optimally to the action of the other. Hoarding while leaving the dog's dumps behind are the best mutual responses – the rational things to do – even when no one wants supply nasty nor sidewalks shortages. It's like prisoners deciding to rat on their accomplice to save their own hide no matter what the co-conspirator decides to do. Even rational prisoners may end up in a Nash equilibrium, where they are worse off than they could have been had they just worked together and kept their mouths shut. Dog dumps and hoarding are variations of this well-known and much studied game called the *Prisoners' Dilemma*.

8.1 Hoarding dog dumps

		Madison	
		Compromise	Get my way
Addison	**Compromise**	3, 3	-2, 6
	Get my way	6, -2	1, 1

If not careful, Addison and Madison may just end up as the prisoners of their own relationship for the same very same reason:

- If Addison is willing to compromise, it's best for Madison to force way.
- If Addison decides to gaslight and bamboozle along to get my way, it's best for Madison to hoodwink too to not to end up with the short end of the stick.
- As before, no matter whether Addison compromises or wants to get away with it, it is best for Madison to boss around.

We try to maximize utility for ourselves but by doing so we may just end up – *all* of us – with a suboptimal result and we could have been in a much better place had we decided to cooperate instead. And that's the problem. Whataboutmeism is about me, and everything so far has been about maximizing individual utility in a continuous competition between players, drivers, partners, lovers, chickenshits dog owners and anxious citizens in a life designed as non-cooperative games.

8.2 Taking down a bigger score – together

There may be something to cooperation rather than competition after all. In the matrix of dog dumps, Madison and arbitrary dog owners do get something out of both picking up (1, 1) – clean surroundings although it is less pleasant on an individual basis to "Pick up" rather than to "Leave it". Similarly, Madison and other citizens do get something out of not hoarding (3, 3) just as Addison and Madison do get something out of compromising ever so often in their relationship. But it is all under the assumption that the parties decide to cooperate. That's a different story and game altogether.

The story and associated game are known as the *stag hunt*. The French philosopher Jean-Jacques Rousseau (1712-1778) came up with the story. It's a tale of two hunters having to decide for themselves to hunt either a hare or a stag and without knowing what the other is going for. What is common knowledge between them however is that one cannot go after stag without the assistance of the other. It is less effort and much less time-consuming chasing down the hare alone, but then again there is not much meat on it. Although minimal risk, minimum effort and time spent but total control and autonomy, brings but a small individual reward. Instead, Rousseau suggests that each hunter, acting individually, is better served going for the more ambitious, but also more rewarding, goal of getting the stag by selling out a bit on the total autonomy enjoyed

8.2 Taking down a bigger score – together

alone yet gain the cooperation of the other hunter and what both hunters brings to the table in terms of added might.

Now the difference between the Prisoner's Dilemma and the stag hunt game is that the former only has one pure-strategy Nash equilibrium in which both "Leave it", "Hoard", "Rat" or try to get your way either way. In the stag hunt there are – not one – but *two* pure-strategy Nash equilibria; one where both players cooperate and one where they don't.

Return to the hoarding example and Madison during COVID-19 lockdown. Either Madison and company can go for the big score (corresponding to stag) and all acquire what is needed now by working together buying normally without crowded conditions and possible virus spreading in the supermarket. Or, opt for the smaller score individually of securing what we personally need (corresponding to hare) in an arms race to get to the stores first, buying as much as possible while assembling *en masse*.

		Other people	
		Buy normally	**Hoard**
Madison	**Buy normally**	3, 3	-2, 2
	Hoard	2, -2	1, 1

If Madison hoards while other people go about their business as usual, Madison just gets a payoff of 2 like more groceries than needed while other parties get zilch and could have saved the trip to the store. Whereas, if all parties buy normally, then everybody get what they need in an orderly manner, 3 each, and if Madison and co. hoard, they might get some, 1 for you, 1 for me, but under stressful and potentially dangerous conditions.

The latter two are the two pure-strategy Nash equilibria. The rational response in the prisoner's dilemma is downright independent of what the other side decides to do, just hoard away everybody! Stag hunt is a on the other hand a coordination game, where everybody is best of doing the same thing – don't hoard when others don't. Thus, the stag hunt has two rational outcomes – if everybody hoards, do it yourself; if nobody hoards, there is no incentive for you to start. In general, if everybody does the same, there is no incentive to deviate so the aggregate behavior is stable. Why not just go for the stag hunt? Because it presupposes trust, something the whataboutmeist doesn't put much stock in.

8.3 Trust me!

It comes down to a question of trust whether the stag hunt can be played together in a manner most beneficial to all parties involved. If Madison believes that her fellow citizens cannot be trusted, and thus assume that

8.3 Trust me!

they will buy more toilet paper and toothpaste than they could possibly carry, then shortage is possibly on the horizon, and it would be rational for Madison to act accordingly and initiate a hoarding spree. Of course, Madison's beliefs about others hoarding could be false and yet still end up in the undesirable equilibrium state of hoarding.

One reason for such a mismatch of beliefs is referred to as a state of *pluralistic ignorance*: Everybody wish to act in a certain way, but simultaneously believes that everybody else acts differently. Believing that everybody will hoard, although no one wants to, may be enough for Madison to start hoarding. Hoarding then becomes a self-fulfilling prophecy. Pluralistic ignorance could also be the cause for leaving the dog shit for your fellow citizen to step in, rat on your partner in crime or always attempting to get your way in your romantic affairs.

Recall from the previous chapter that exactly trust, from the point of view of whataboutmeism, is considered overrated and not to be pursued: It paves the way for vulnerability and insecurity; trust is in conflict with both the imperative and formula of whataboutmeism and finally; others are assumed somewhat incompetent in defining the common good using the secular theodicy argument to leverage your right to define what the common good is. The consequence of this stance is viewing life as a string of (more or less) zero-sum deals, endless non-cooperative games all the way to the war of attrition. Prisoner's dilemmas here, there and everywhere. And even if we do decide to cooperate in the stag hunt, there is still a pure-strategy Nash equilibrium where we "Leave it", "Hoard", "Rat" and "Get my way" when we don't trust each other.

To get to the desirable pure-strategy Nash equilibrium in the stag hunt of gaining more together than we could apart, trust is the trick to cooperation. Keep cool in times of trouble and commit to each other not getting carried away in ego when the s*** hits the fan. Scottish Enlightenment, philosopher, economist, librarian and essayist, David Hume (1711-1776), had a way of always wrapping precision in poetry pre-empting the travesty of common obstruction and lack of trust while pointing to important ingredients of whataboutmeism:

> Your corn is ripe today; mine will be so tomorrow. 'Tis profitable for us both, that I should labour with you today, and that you should aid me tomorrow. I have no kindness for you, and know you have as little for me. I will not, therefore, take any pains upon your account; and should I labour with you upon my own account, in expectation of a return, I know I should be disappointed, and that I should in vain

8.3 Trust me!

> depend upon your gratitude. Here then I leave you to labour alone; You treat me in the same manner. The seasons change; and both of us lose our harvests for want of mutual confidence and security.[17]

Think of Addison and Madison's relationship as a proxy for our own relation when whataboutmeism rears its ugly head with either one of the parties or both. Then aim for:

- Converting the non-cooperative prisoner's dilemma into a cooperative game of stag hunt.
- That means converging on a common good *together* and *compromise* ever so often.
- Entrust each other with the belief that you are in this together for the common good mutually agreed on and nobody knows better than you two!

Then we will mutually get the highest payoffs (3, 3) by trusting each other, working together while not being neither a coward nor a crook. Being systematically dead set on getting your way in the spirit of whataboutmeism will get us a lower mutual payoff, 1 for you, 1 for me with little to no trust fueling the egoism of perversion.

		Addison	
		Compromise	**Get my way**
Madison	**Compromise**	**3, 3**	-2, 2
	Get my way	2, -2	**1, 1**

It may sound as if you just must give yourself up and continuously compromise or give to do the right thing. Wrong! Individual autonomy is not just cashed out in perverted egoism and whataboutmeism alone. Sometimes it is just not about you, and even when it is, strike a balance not systematically blaming others if it fails – you are in it too. It's about us!

Chapter 9
What about you? It's about us!

Individual autonomy is not to be confused with whataboutmeism. Autonomy is about individual authority, enlightened decision-making, rationality, fair and balanced reasoning about yourself and the ones around you. It's about trust, being good people, in a group of other autonomous people lining up for stag hunt in the interest of collective well-being and democracy. Egoism is not autonomy, groupthink not democracy.

9.1 Back to birthdays

My father's harangue as I was complaining about not trying out my RC car at my 12-year birthday party served more than one purpose. To rehearse: I asked, "What about me" and my father replied:

> What about you?! This is not about you! It's about them! You get to play with it all the time. Even if you are not to race it whenever you fancy, cherish the fun your friends are having trying out the RC-car. Oh, and do yourself a favor: Cherish their fun without thinking that you somehow missed out on

something along the way just because your friends tried *your* car!

Besides encouraging me not to view the situation as a zero-sum game, he is also giving me the recipe for *social success*. Odds would be stacked against me of my friends returning on other occasions if they were to realize that I was exclusively worrying about what I stood to gain personally in their presence. If I'm able to cherish and enjoy their pleasure while playing an active role of letting them play with my toys too without one spreadsheet to rule them all for which I keep the scoreboard, chances are my friends will return on a later date.

My father was telling me to get it straight. A birthday gathering is about celebrating *you together*, but it is not intrinsically about you. It's about everybody present and the party as such is there to exactly ensure the successful collective outcome. At my recent 50th birthday party, almost 40 years down the line, I followed this blueprint of making the party be about all of us; food, drinks and dance for everybody to enjoy and take *collective ownership of my party*. And I'm pretty sure they will all return for my 60th birthday event. It is a short-sighted social strategy to making it all about you. Either you are a risk averse coward or a risk seeking crook either way trying to *control* your social

9.1 Back to birthdays

surroundings. Eventually people will have had enough – and they will stay clear of you.

We don't know the fate of Addison and Madison and their relationship if for no other reason because they are fictitious but hopefully characters, we may relate to one way or the other. Chances are, on the current trajectory, eventually they will have to call it quits due to lack of cooperation between compromising and getting their way. We might sometimes believe that *we are cooperating when what we are really doing is negotiating*. Addison and Madison are continuously negotiating rather than cooperating because that's the nature of the relationship when set up as a game of chicken.

Cooperation and negotiation are not the same thing. Once mutually satisfactory negotiations have come to conclusion, then parties may commence cooperating towards some goal mutually agreed upon. But negotiation by itself is not cooperation and negotiations may go either way. A negotiation in gridlock is not cooperation in and by itself, although some cooperative effort is likely required to get out of the predicament if possible. Designing life as a zero-sum game may often enough confuse cooperation with negotiation to secure my way among friends, family members, colleagues, markets of investment, the imaginative relationship between Addison and Madison and the real deal.

9.2 Love is not a zero-sum game

Throughout a lifetime we date a score of different people. Suppose there is an equal probability of winning and losing in this dating game of life. Assuming dating some indefinite number of people, the expected gains and losses will add up to zero. Thus, why date at all in the first place, or better still, why not just be indifferent to dating? The payoff is equivalent to when not dating. Worse than that, better be safe and decide against dating all together. Winning or losing the same amount, prospect theory conveyed that losses hurt so much more than the equivalent potential gains.

In this dating game with equal probabilities of winning and losing with potential losses hurting so much more than the potential gains, the probabilistic emotional state just looks so bad, so why play the game in the first place? Of course, if I expect that my personal probability of winning is greater than 50%, then it makes sense dating. But then again, if I meet a person who only dates when the numbers are stacked in the same way of more than 50% chances of winning, then we will self-select and date only when there is a more than 50% chance of a score. Playing only weak opponents consistently would be the exception, but that just means that their expertise in playing the dating game is less impressive than mine. With the zero-sum game having long run negative payoffs, as witnessed in Addison and Madison's

9.2 Love is not a zero-sum game

war of attrition, there is no incentive to play the dating game period.

But we play it anyway again and again, in part because of a bias making us believe that there is a higher probability of winning making it worthwhile at least for a while. After some period of gameplay and experience, we will likely conclude that the zero-sum game is not worth it anyway. But and this is the *big but*, dating may lead to something else not factored in so far. Namely love. And although dating may be a zero-sum game, love is not. With two people in love, their jointly maximized utility exceeds the sum of their individual utilities counting from when they were on their own. If a happy Addison falls in love with a happy Madison, then they'll be super happy together which is not a zero-sum game, but the situation where everybody gains. Keep playing the zero-sum game of dating and with any luck you and your partner of choice will eventually end up in the stag hunt based on trust and cooperation. What about me? This is not about you! It's about us!

And finally remember: Just because I spend time with my father after the divorce, doesn't mean I don't love my mother. And just because Addison spends time with a child of a previous relation doesn't mean Addison loves Madison any less. They are just items on separate spreadsheets with no joint scoreboard to keep. Issuing speech acts like "I expect that you …", "I'll be disappointed if you …", "I would really wish that you …" while designing life as a zero-sum game is to count your chickens before they hatch. The party in question may either not be able, nor even want, to cash in whatever it is you are asking for because your love doesn't view love as a zero-sum game in the first place, so why should you when you can do better together? Actor Julianne Moore hits it right on the head when she proclaimed:

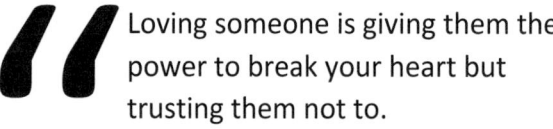

> Loving someone is giving them the power to break your heart but trusting them not to.

9.3 Consumer democracy

A master's student came to my office the other day for our first meeting about the thesis to conclude master's program. First things first, the student sits down and proclaims that the thesis is to be written on Mauritius, an exotic Indian Ocean island nation, the trip to commence immediately. I answer: "That sounds good, seems to me you have to go see a travel agent rather than a professor of philosophy." The university doesn't have a complimentary travel service for students run by faculty members. The order of business is master's student first, and a trip to Mauritius, for largely personal reasons, a distant second. The student initially saw it the other way around thinking what about me. But the university is not installed to stimulate some personal gain, but secure best practice science and as a civil society actor add to the body of human knowledge benefitting us all. Ask instead how you can chip in, and if a trip to Mauritius is a means to this common end of beneficial knowledge aggregation, then by any means necessary.

Similarly, my friend Bjarke, who works for the municipality school services, was called to his office where two well-to-do parents were awaiting his coming. As Bjarke sat down the couple asked what he could do for their kid. Bjarke asked if something was wrong as to their child's situation in school, to which they answered: "No, not at all, things are great, we just wanted to come by and check whether there is something the municipality can do for our family. Sort of a 360 degrees check-up making sure we're not missing out on something the municipality has to offer!" Same thing again, what about me in an institutional setting. It's not about you, it's about us, and you are not necessarily suspiciously missing out on anything, especially if things, by and large, admittedly are going well. The American scholar, author and organizational scholar Warren G. Bennis (1925-2014) sums it up adequately when he said:

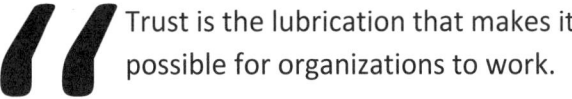

> Trust is the lubrication that makes it possible for organizations to work.

We can blame family, friends, foes, colleagues, employers, BigTech, the political climate, the stock market and other road users – whoever and whatever for our busted expectations and personal mischief, only to realize that none of the above would be in business if it weren't because there are markets for them in which we more often than not individually partake. Some it is also on us.

American philosopher, abolitionist and essayist, Ralph Waldo Emerson (1803-1882) once said:

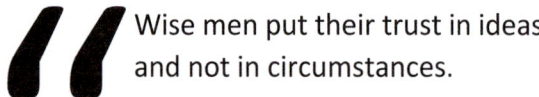

> Wise men put their trust in ideas and not in circumstances.

A central philosophical idea to which we time and time again pay homage to is to claim of individual autonomy, integrity and the right to self-government. That's not equivalent to psychological or ethical egoism nor whataboutmeism. In Kant's realm of thought, autonomy is a particular capacity only humans possess to act in accordance with an objective morality based in no small part on his categorical imperative and the formula of humanity. With Emerson in mind, whataboutmeism is less about ideas, but indeed very *circumstantial* by

- **designing life as a whataboutme game (towards partners, friends, family members, colleagues etc.)**

9.3 Consumer democracy

- **flipping the Kantian categorical imperative and formula of humanism on their heads,**
- **supplementing the lopsided principles with perverted egoism and**
- **gaslighting the social surroundings into the belief that the secular theodicy argument of knowing best what benefits everyone**
- **entailing both a sense of torchbearer entitlement and**
- **reactionary reproduction of what's around already while**
- **being either a risk averse coward or a risk-seeking crook, which either way**
- **supposedly exonerates the protagonist of any moral responsibility or accountability but places such squarely on the social environs.**

The price to be paid for no accountability is no trust as yet another Enlightenment notability, English-born American political theorist, activist and revolutionary, Thomas Paine (1737-1809) proclaims:

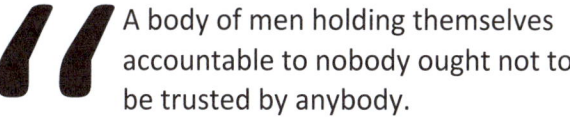

A body of men holding themselves accountable to nobody ought not to be trusted by anybody.

Autonomy, integrity and accountability received much attention during the Enlightenment and for good reason too. Individual autonomy – the ability to reason out the leads, weigh pros and cons, deliberate, decide and act rationally human by human – was then, and still is, the central component or building block for realizing the emerging democracies in the US and around Europe back in the day of Enlightenment. Democracy is not merely about having a vote, a robust democracy is about having an informed vote, and that's a different story all together and includes important items like trust, cooperation and collective maximization of utility.

Preferring democracy as the best way to configure and rule a state is basically a position, a belief or a viewpoint but it doesn't make everything that takes place in a democracy a stance, belief or subjective point of view. From a democracy, a collection of facts follows concerning everything from the tripartition of power and the parliament of representatives, to a market- or centrally planned economy and so forth. Several institutions are installed to safeguard, regulate and take part in optimizing the democratic structure that one has attitudinally been able to agree on.

More specifically a line of institutions is installed that amount to democratic indicators, gatekeepers and guardians – checks and balances through and through to install and ensure trust. Public officials who investigate citizens' complaints against the government come to mind, along with various other investigative committees, national auditors and so forth. Their job generally consists in uncovering, collecting, analysing

9.3 Consumer democracy

and assessing the facts or consequences which the democratic form of government occasions — from legal complaints to accusations of fraud, corruption etc. These institutions must meet stringent criteria of impartiality and often even scientific probity precisely to ensure that what follows from the chosen form of government isn't just something one can doubt at random, disregard or archive in the waste-bin as advertisements for various political products. Among politicians there is a widespread tendency to appoint fact-finding commissions, investigative committees and so on. The mandate of these commissions and committees is to produce and summarize knowledge and evidence as the basis for political deliberation, decision and action. Often enough, at the same rate that these commissions have delivered their findings politicians have routinely questioned the results brought forth: Especially when they haven't proven advantageous for a certain political agenda.

Luckily one may make objections both of a political nature and pertaining to the quality of the facts the democratic institutions may base their actions upon. The latter however requires that one can produce better results or facts. By contrast, it's completely free of charge and contrary to the democratic form of government to comfortably claim that those institutions which uncover unpopular facts are merely political agents or plainly mistaken despite their composition and intended impartiality. Roughly speaking, if we waive the respect for facts and truth, we waive the respect for real democracy. Facts and truth, or in short, *true information*, is a necessary, but not sufficient condition for knowledge, and for democracy.

The American political scientist, economist and celebrated author, Francis Fukuyama, points to the fact that corruption is a lurking danger in every government body because of the human inclination to favour him/herself, family and friends.[18] In the fight against corruption, the establishment of trust and the construction of communal institutions is the first law of building a state. A fundamental premise is to understand how democracy works as social interaction and coordination. The coordination game is often played by having a pretty good idea of the strategies of other players to either act in concert with or block them to maximize personal utility — or by realizing that the greatest payoff for everybody is realized by cooperation and coordination.

The latter is the lesson from the Prisoner's Dilemma: Every citizen realizes that the aim of maximizing personal utility may facilitate forms of government which are democratically questionable and often lead to the oppression of other individuals and groups. That goes for everybody below the top of the oligarchy or regent. Such a position is naturally reserved for the

9.3 Consumer democracy

select few, which may explain why democracy is to be preferred over and above the other rules of government. By cooperation and coordination, the utility will be greater for all – and that's the essential message of the stag hunt game assuming success only via trust.

There is once more a price to pay. Everybody must give up the same amount of sovereignty to the societal institutions. In return, the managers of power are obligated to carefully and correctly respect the common rules set forth. Democracy is really the answer to a challenging coordination problem. The ability to institutionalize norms of trust, objectivity and impartiality are the beacons of democracy. Without these beacons democratic coordination is impossible. Thus when players, especially those who define the rules of the game such as government heads, spin doctors and public officials begin perfunctory practice with the truth and exercise corruption then that would be like letting a lion loose among the zebras. Abuse of power may lead to systemic distrust, which in turn may spread like a cascade through the societal system.

Whataboutmeism doesn't square with democracy in the sense just outlined above. It may rhyme somewhat with a variant descent from higher principles of proper politics: *Consumer democracy*. Politics are put up for sale like commercial products after which citizens are invited to judge them accordingly and possibly buy into them as well. Whereas politics is often taken to be contemplated, lived and judged relative to standards of the public good, purchase decisions are typically not held to such noble demands. Buying something for *me* may very well be divested of the idea of the public; a private or individual affair and once present in the political arena it could indeed be corroding the very principles of the public good. Voting becomes like going to the supermarket walking though aisles of political products and finding the offers best suited for me my preferences and what then the "public good" is for me. Whether or not some political product ends up in the shopping cart possibly under fire sale depends on whether it satisfactorily answers one question: What about me? The check-out counter becomes the voting booth where the ballot is cast, or the political investment made. Then it just remains to be seen whether the politicians can cash in your expectations, if not, it is all on them again politically and morally. If citizens at large buy into consumer democracy we may as well forget about putting trust in ideas rather than circumstances, autonomy, integrity, accountability, higher principles of politics and the public good.

9.4 Mighty or moronic

Probably no one person would voluntarily want a subscription to whataboutmeism. Having said that, situations do exist in which it is completely reasonable

9.4 Mighty or moronic

that it is about oneself, the group, democracy. Obvious injustice, wrongful discrimination, arbitrary persecution, disgusting harassment, against individuals, groups, society and democracy have indeed been committed time and again. In such cases, it is also about oneself. One may reasonably ask: "What about me, my group, my society or the democracy I live in?"

What happens in the long run if the whataboutme game is repeated over and over again? One possible outcome is that the whataboutmeist experiences the environment punishes him or her, since the ones around receive lower utility. The whataboutme player might choose to play a different strategy just to avoid punishment. The whataboutmeist in a sense shows consideration and does something generally viewed as "good", because it ultimately also benefits the person in question. Whataboutmeism then becomes ethical egoism. This sort of selfishness may in fact also benefit of others as already noted in chapter 3.

But a full-fledged whataboutmeist views everything from a vantage point of self-interest without a care in the world for altruistic derivatives that possibly also could be beneficial to the individual. A protégé of whataboutmeism tries to take advantage of the environment by presenting things in such a way that it seems as if the gameplay (potentially) different from the game being played. Suppose I am now the one playing whataboutme. I thus try to assign lower utility to others than what they are entitled to. My point of departure is precisely that I am the most important and simultaneously know what is best for everyone given the theodicy argument. I may just end up playing furiously irrational over time as I end up getting punished for playing the game as a moron. So, it shouldn't be worth it to me. Turns out, I will not even be a homo economicus trying to maximize utility for myself, but rather an *irrational moronicus* in the long run.

Another argument for not falling pray of whataboutmeism is hedonic adaption. The utility of best outcome decreases over time if you generate the best outcome all the time. It becomes a habit and loses its magic. For that reason alone, it is not necessarily rational to always try to get it your way. It feels better to get your way when you don't always get it. In addition, one's utility value assignment is also a function of uncertainty. The gain of catching a train is greatly affected by whether you arrived well ahead of time or made it just in time. If you caught the train last minute, you get a huge emotional reward (especially if you're on route to something important). Precisely due to the uncertainty of whether you were going to make it.

In the rigged whataboutme game, it will apply that if you know that the other person will always back off, the utility value is less than if there was an element of

9.4 Mighty or moronic

chance or risk. The disadvantage of tyranny is that you don't take chances or vary your actions. In the long run this will grow to a much worse outcome. Like in chess. It is increasingly boring to play against someone you constantly know you will lose to, or for that matter, always win against. Lower utility in both cases in the long run.

My good friend, Lise Nørgaard, said of herself that she always played in major by the end. Too much music in minor, she couldn't help but laughing at some point. If are to end up playing in major, remember there is a positive variant, where witnessing the happiness of others means that you start expecting more of yourself. Seeing the success of others may motivate you to push harder on your own. Instead of thinking that the world is unfair because the same thing doesn't automatically happen to you, you instead start thinking of making an effort, let oneself inspire and be open-mindedly curious about others. Such a vantage point may strengthen collaborations, communities, trust, democratic citizenship, security the well-being of oneself as well as others. Social comparison may have perfectly benign consequences motivating us to become more like those we consider living life in an independent, responsible and trustful way. That's as far from whataboutmeism as you may possibly get, but as close to what we as humans genuinely want to be.

Acknowledgements

This has been some fun, enlightening but also a self-examining and self-critical experience to write this book. Once I set fingers to keyboard, it just came out chapter by chapter. It must have been rumbling around in my head ever since I was twelve and my father taught me first lessons of whataboutmeism at my birthday party.

I couldn't have done it without the help of a selective few. My great old friend and designated consultant on this project, Bjarke Malmstrøm Jensen, put his twisted mind to work, came up with telling exemplars, had pointy observations and razor-sharp reasoning all the way through. My other friend, Teit Molter, was always there to listen and laugh and took care of uncountable backups. Professor Gregory Wheeler, probability theorist and behavioral economists, long-time colleague and friend suggested many valuable angles of approach and supplied significant examples and cases from a long life in academia. The same goes for Professor Kevin Zollman of Carnegie Mellon University who also set me straight with his brilliant knowledge of game theory, decision theory and formal epistemology. Mads Vestergaard, PhD in Philosophy, co-founder, back in the day, of the now dissolved Nihilistic Folk Party of Denmark, techno-DJ, accomplished author and quite off the normal distribution in ways more than one, also chipped in with noble alterations and nutty installments to earlier drafts of the manuscript.

I would like to thank my editor Christopher Wilby for his great help and enthusiastic support all the way to publication. I'm grateful to Ties Nijssen, Floor Oosting, Palani Murugesan and Springer Nature for taking on *Whataboutmeism*.

The family was also along for the ride. My mother, Annette Møller, came up with stockpiles of examples from lived life. The pride of my life, my son, Milton W. Hendricks, has experienced what whataboutmeism is like, and has provided extremely valuable insights throughout. Camilla Mehlsen, the love of my life has stood by me all along, and put her curly brains to work on ideas, texts and edits. Last, but not least, there is my father, Elbert L. Hendricks, without whom this book never would have been conceived.

Notes

¹ For a brilliant introduction to game theory in general and how to raise your children in particular, refer to Raeburn, P. & Zollman, K. (2016). *The Game Theorist's Guide to Parenting: How the Science of Strategic Thinking Can Help You Deal with the Toughest Negotiators You Know -Your Kids*. New York: Scientific American: Farrar, Strauss and Giroux.

² OECD Insights: Human Capital (2014). "What is social capital?", verified 24.07.2022: http://www.oecd.org/insights/37966934.pdf

³ Kierkegaard, S. (1843). *Either/Or*, Part II. Edited and Translated by Hong, H.V. & Hong, E.H. Princeton: Princeton University Press, 2013: 159.

⁴ United Nations: Universal Declaration of Human Rights, verified 22.07.2022: https://www.un.org/en/about-us/universal-declaration-of-human-rights

⁵ Kant, I. (1785), *Groundwork of the Metaphysic of Morals*. Cambridge University Press; 2nd edition (May 21, 2012).

⁶ *Ibid*.

⁷ Bentham, J. (1832). *Constitutional Code*. Oxford: Oxford University Press, 1987: Introduction, §2.

8 *Theodicy* by Freiherr von Gottfried Wilhelm Leibniz, verified June 10, 2017: http://www.gutenberg.org/files/17147/17147-h/17147-h.htm

9 At a fundraiser in New York on September 9, 2016 during the election campaign, Hillary Clinton said as follows: "You know, to just be grossly generalistic, you could put half of Trump's

supporters into what I call the basket of deplorables. Right? They're racist, sexist, homophobic, xenophobic — Islamophobic — you name it."

[10] Mankiw, N.G. (2010). "Spreading the Wealth Around: Reflections Inspired by Joe the Plumber", *National Bureau of Economic Research*, working paper 15846, verified 06.08.2022: http://www.nber.org/papers/w15846.pdf

[11] *The Silence of the Lambs* (1991). Anthony Hopkins: Dr. Hannibal Lecter. Quotes, IMDb: https://www.imdb.com/title/tt0102926/characters/nm0000164

[12] Moynihan, D.P. (1983). Column, *The Washington Post*, January 18.

[13] Cruz, J. (2019). "Humble Trust", *Philosophical Studies*, 176(4): 933-953.

[14] The seminal papers on prospect theory: Kahneman, D. & Tversky, A. (1979). "Prospect theory: An analysis of decision under risk." *Econometrica, 47*, 263-291; Tversky, A. & Kahneman, D. (1992). "Advances in prospect theory: Cumulative representation of uncertainty", *Journal of Risk and Uncertainty*, 5: 297-323; and the later accessible book: Kahneman, D. (2011). *Thinking, fast and slow.* London: Allen Lane.

[15] The seminal papers on prospect theory: Kahneman, D. & Tversky, A. (1979). "Prospect theory: An analysis of decision under risk." *Econometrica, 47*, 263-291; Tversky, A. & Kahneman, D. (1992). "Advances in prospect theory: Cumulative representation of uncertainty", *Journal of Risk and Uncertainty*, 5: 297-323; and the later accessible book: Kahneman, D. (2011). *Thinking, fast and slow.* London: Allen Lane.

[16] Kahneman, D. (2011). *Thinking, fast and slow.* London: Allen Lane: 242.

[17] Hume, D. (1965). *A Treatise of Human Nature*. (Ed.) Selby-Bigge: (p. III.II.v). Oxford: Oxford University Press. Originally published 1739.

[18] Fukuyama, F. (2012). *The Origins of Political Order.* New York: Farrar, Strauss and Giroux.

MIX
Papier aus verantwortungsvollen Quellen
Paper from responsible sources
FSC® C105338

If you have any concerns about our products,
you can contact us on
ProductSafety@springernature.com

In case Publisher is established outside the EU,
the EU authorized representative is:
**Springer Nature Customer Service Center GmbH
Europaplatz 3, 69115 Heidelberg, Germany**

Printed by Libri Plureos GmbH
in Hamburg, Germany